Lecture Notes in Economics and Mathematical Systems

475

Springer-Verlag Berlin Heidelberg GmbH

Leo Kaas

Dynamic Macroeconomics with Imperfect Competition

Springer

Author

Dr. Leo Kaas
Institute for Advanced Studies
Department of Economics
Stumpergasse 56
A-1060 Vienna- Austria

Library of Congress Cataloging-in-Publication Data

Kaas, Leo, 1969-
Dynamic macroeconomics with imperfect competition / Leo Kaas.
p. cm. -- (Lecture notes in economics and mathematica systems ; 475)
Includes bibliographical references (p.).
ISBN 978-3-540-66029-3 ISBN 978-3-642-58479-4 (eBook)
DOI 10.1007/978-3-642-58479-4
1. Competition, Imperfect--Mathematical models.
2. Macroeconomics--Mathematical models. 3. Equilibrium (Economics)--Mathematical models. I. Title. II. Series.
HB238.K33 1999
338.6'048--dc21 99-34450
CIP

ISSN 0075-8442
ISBN 978-3-540-66029-3

Typesetting: Camera ready by author
SPIN: 10699851 42/3143-543210 - Printed on acid-free paper

To my parents

Preface

This thesis was stimulated throughout the time of my participation in a research project on Dynamic Macroeconomics, supported by the German Research Foundation (DFG). The starting point was the central question of how to integrate price setting firms in a dynamic disequilibrium model. Almost all recent literature on imperfect competition in macroeconomics applies the *objective demand* approach by assuming that firms know the true demand curve they are faced with. While this approach can be applied in temporary monetary equilibrium models, it proves inadequate for formulating price adjustment in a dynamic disequilibrium model, where it has to be replaced by the concept of *subjective demand*. Based on this distinction, the thesis starts out with a comparison of the concepts of subjective and objective demand in an abstract framework and surveys the literature on general equilibrium theory with imperfect competition. The objective demand approach is criticized not only on the grounds of its strong rationality requirements and existence problems, but also by the observation that it cannot be applied successfully to characterize determinate rational expectations equilibria in intertemporal macroeconomics. Finally, price setting firms using subjective demand functions are integrated in a dynamic disequilibrium model in order to study monopolistic and oligopolistic price adjustment.

In May 1998, this thesis was submitted as a dissertation at the Department of Economics of the University of Bielefeld. At this point, I would like to thank both my supervisors, Prof. Volker Böhm and Prof. Thorsten Hens, for many helpful discussions, stimulations and their criticisms throughout all stages of research and completion. I am particularly grateful to Prof. Paul Madden, Prof. Neil Rankin, and Prof. Claus Weddepohl for valuable comments on previous working paper versions. I am also very grateful to Prof. Gerd Weinrich for helpful comments on this thesis. Parts of the thesis have been presented at workshops and conferences in Berne, Bielefeld, Oxford, Toulouse, and Vienna, whose participants I would like to thank for their feedback. Finally, I thank my former colleagues Nicole Deutscher, Eckart Jäger, Jens-Ulrich Peter, and Dr. Matthias Raith for their careful proof-reading and for useful remarks, and Uli Middelberg for his indispensable computer support. Of course, all remaining errors are my own.

Vienna, October 1998

Contents

Chapter 1

Introduction

> "*Clearly, product markets, as well as labor markets, should be modeled as imperfectly competitive.*"
>
> J. Tobin (1993)

Nowadays almost all of macroeconomic theory is based on microeconomic foundations. It is the widespread opinion that a full understanding and explanation of macroeconomic phenomena like inflation, unemployment, growth, and the business cycle require a model in which economic decision units and their interaction are described explicitly. Only models with a microeconomic foundation are immune to criticism concerning "ad hoc"–assumptions of macroeconomic relationships, and only these models allow an assessment of welfare effects of policy measures.

In the 1970's, the development of macroeconomics with microeconomic foundations followed two important branches. On the one hand, Neo–Keynesian macroeconomic models[1] were introduced by Barro & Grossman (1971), Benassy (1975), and Malinvaud (1977). They follow the

[1] The term "Neo–Keynesian" was introduced by Benassy (1975). Other expressions are "disequilibrium theory" or "fixprice theory".

idea that prices are fixed in the short–run, and that trade takes place at non–Walrasian prices. The major advantage of this approach is first, that it provides a description of the concepts of the Keynesian effective demand and the multiplier in a consistent microeconomic framework, and second, that it allows a classification of different types of unemployment, Keynesian or Classical. On the other hand, New Classical models (e.g. Lucas (1972), and later the RBC literature) assume that prices in competitive markets adjust continually to equilibrate supply and demand. A fundamental principle is the hypothesis of rational expectations which claims that agents' forecasts of their future economic environment are not subject to systematic errors. According to this view, persistent unemployment is a structural phenomenon which cannot be fought by economic policy. For a detailed and critical exposition of both directions of research, refer to Felderer & Homburg (1987), for example.

In the past two decades, important efforts have been made to improve and to bring together these approaches of macroeconomic theory. Most of the relevant literature, commonly summarized under the heading of "New Keynesian macroeconomics,"[2] follows the New Classical hypothesis of rational expectations, but deviates in one way or another from the paradigm of market efficiency by assuming market imperfections of one form or another.[3] The most important imperfection is the assumption of imperfect competition in labor or output markets.[4]

There were early studies incorporating imperfectly competitive agents in a macroeconomic model by Benassy (1977) and Negishi (1979). Their models adopted the so–called "subjective demand approach" which

[2]Dixon & Rankin (1995) prefer the expression "New Macroeconomics", since these models need not necessarily have Keynesian features.

[3]Mankiw (1990) goes so far as to claim that "*the axiom of rational expectations is as firmly established in economic methodology as the axioms that firms maximize profit and households maximize utility.*"

[4]Others are implicit wage contracts, efficiency wages, or credit market imperfections. For surveys and references, see Mankiw & Romer (1991), Mankiw (1990), and Dixon & Rankin (1995).

claims that price–setting agents have a subjective perception of their demand curve which passes through the equilibrium price/quantity–pair but is arbitrary beyond the equilibrium. Since this high degree of arbitrariness leads to a large number of possible equilibrium allocations, and since subjective demand perceptions have been criticized for not being "perfectly rational", the whole subjective demand approach has been abandoned. A breakthrough has been achieved by various studies applying the "objective demand approach" in macroeconomic general equilibrium models. The basic idea is that imperfectly competitive agents know the "true" demand curve of the competitive sector of the economy. The articles of Hart (1982) and Blanchard & Kiyotaki (1987) have been most influential in this direction. They contain static models incorporating money[5] as argument in the utility function, which describe equilibrium situations where firms are imperfect competitors on the output market, while unions or workers are imperfect competitors on the labor market. Clearly, such equilibria are typically inefficient with an output level below the perfectly competitive one, and they may display involuntary unemployment. These models share some Keynesian as well as Classical features in the sense that fiscal policy can raise the output, while monetary policy is neutral, as long as real money balances enter the utility function.[6]

More recent work has focused on intertemporal macroeconomics with imperfect competition. In some models, price setters face price adjustment costs and have to decide on the optimal timing and level of price changes (Ball & Romer (1989)). Other models incorporate imperfect competition in OLG models (Schultz (1992)), RBC models (Rotemberg & Woodford (1992)), and in models of endogenous growth (Romer (1990)). All of these have in common that they restrict attention to rational expectations equilibria.

[5]Hart refers to a "nonproduced good" instead of "money".

[6]This corresponds to unit–elastic price expectations in an underlying temporary equilibrium model.

Dixon & Rankin (1995) emphasize the "*coherence of the New Macroeconomics,*" and indeed it appears to be one of the major contributions of microeconomic foundations to have established a broad consensus among macroeconomic theorists. However, there are two important objections against the recent contributions to intertemporal macroeconomics with imperfect competition.

First, there is an objection from a purely microeconomic perspective concerning static as well as dynamic models.[7] Most of the work mentioned above is confined to specific parameterizations of technologies and preferences. While this may be a useful guideline for didactic purposes, it must be emphasized that a general theory - even for the simplest types of macroeconomic models - cannot be developed since an equilibrium with objective demand need not exist.[8] Since the early 1960s, much effort has been undertaken to extend the Arrow–Debreu general equilibrium model to imperfect competition. However, various complications in these models led Bonanno (1990) to emphasize that "*unfortunately we are still far from a satisfactory theory of general equilibrium with imperfect competition.*" At any rate, the whole theory of macroeconomics with imperfect competition is very wobbly on its microeconomic legs. One might argue that such objections are irrelevant for macroeconomists who are primarily interested in simple models suited for econometric tests.[9] But from a theoretical perspective the whole microeconomic "foundation" must be questioned if it rests on such specific assumptions.

---

[7]Another important microeconomic objection, which concerns almost all of macroeconomics with microeconomic foundations, but which will not be discussed further in this thesis, can be raised against the "representative agent approach." For an exposition of this problem, see Kirman (1992).

[8]Hart (1982) is well aware of these problems when he remarks that "*generalizing the model could be hard. For this reason, the analysis presented here should be considered more as an extended example than as a general model.*"

[9]Such an argument corresponds to Friedman's (1953) methodological position which judges economic models by its predictions instead by the realism and generality of its assumptions.

Second, almost all of the literature on intertemporal macroeconomics with imperfect competition restricts the analysis to rational expectations equilibria. One of the most popular arguments in favor of the assumption of rational expectations is the following: if the implications of certain imperfections in the economy were to be analyzed, one should keep the model free from other imperfections like non–rational expectations, for instance. Thus, rational expectations are, even though often unrealistic, understood as an idealization of the expectation formation processes of the economic agents. Nevertheless, it must be emphasized that rational expectations should appear as a special case of a dynamic economic process in which agents form expectations on the basis of the past.[10] In such a framework it should be possible to study the dynamics under learning of "boundedly rational" agents, and it should be verifiable whether such dynamics converge to a rational expectations equilibrium or not. However, it is not at all clear how such processes can be formulated in these models.

There are different possibilities to frame a dynamic macroeconomic model. The most obvious one to most economists is to apply the temporary equilibrium method as introduced by Hicks (1946) and developed in full rigor by Grandmont (1983). The general idea of this approach is to assume a dynamic economy where trade takes place sequentially in periods ("weeks" in the exposition of Hicks). At the beginning of each period (on Hicks' "Monday"), an adjustment in prices or quantities converges to a temporary equilibrium, in which the prevailing prices and trade contracts are determined for the rest of the period. This process of adjustment to the temporary equilibrium is not modeled explicitly, and it is assumed to take place infinitely fast. In other words, a temporary equilibrium is established immediately in each period. Agents form expectations using only information from the past, so that expectations need not be fulfilled.

[10] A general setting, in which this problem can be formulated, has been introduced by Böhm & Wenzelburger (1995).

The special case where expectations are realized may be called *intertemporal equilibrium* or, in Hicks' notion, *equilibrium over time.* Given a specific forecast rule of the agents, sequences of temporary equilibria can be generated as solutions of a well–defined dynamical system only if the existence of a temporary equilibrium is guaranteed *for all* variables predetermined in the past. Since it is also desirable to formulate such a dynamic process under different specifications of the forecast behavior, existence of temporary equilibria has moreover to be guaranteed for *arbitrary* specifications of the agents' forecast rules.

In the case where all agents are price–takers, Grandmont (1982) distinguishes between two main temporary equilibrium concepts. On the one hand, there is a temporary Walrasian equilibrium, in which prices are flexible and a competitive equilibrium on all spot markets is established in each period. The main weakness of this equilibrium is that its existence depends on the way agents form their expectations. More precisely, only if price expectations are bounded above and below can existence be guaranteed (see Grandmont (1983)). On the other hand, in a temporary equilibrium with quantity rationing, prices are fixed and trade takes place at a fixprice equilibrium, as defined by Drèze (1975) and Benassy (1975). While here existence of equilibria can be proven for all strictly positive price vectors and for arbitrary continuous forecast functions (cf. Benassy (1982), Chapter 8), these models have been criticized since the assumption of price rigidity and the formation of prices lack a microeconomic foundation.[11]

Under imperfect competition, a temporary equilibrium can be defined as an equilibrium with objective demand in a single period.[12] Temporary equilibrium models, which are related to the static model of Hart (1982), have been formulated by Dehez (1985), D'Aspremont, Dos Santos Fer-

[11] Of course, the second criticism affects also the concept of a competitive equilibrium. See the critique of Arrow (1959).

[12] A temporary equilibrium with subjective demand has been defined by Benassy (1982), Chapter 9.

reira & Gérard-Varet (1989) and Rankin (1992). An important feature of such models is that a characterization of intertemporal equilibria leads to unavoidable problems of indeterminacy. In particular, it turns out that there is typically a continuum of stationary equilibria. Such results have first been established by Rankin (1992) and they will be further elaborated in this thesis. Because of this indeterminacy, there is a particular need for a selection among these equilibria by learning dynamics. However, these dynamics can only be formulated if temporary equilibria exist for some specific learning scheme of the agents. As pointed out above, existence of an equilibrium with objective demand is problematic even in the static case, and one can expect that the situation gets worse in the temporary equilibrium case where expectation effects also play an important role.[13] Thus, the approach of temporary equilibria with objective demand does not seem to be appropriate for the study of dynamic macroeconomics with imperfect competition.

In order to describe a globally defined dynamic process, it seems necessary to abandon the idea that in each period an infinitely fast adjustment to an equilibrium takes place, and instead to formulate such an adjustment process explicitly. There are many different ways to model the adjustment process, as there are many different concepts of equilibrium with imperfect competition. In any case, an equilibrium should appear as a stationary state if we presume a more or less plausible learning process of the agents. For example, one may think of a process, in which firms choose their output level in advance and let an auctioneer determine the market clearing prices. If the observed price differs from a firm's expected price, it will alter its demand expectations. Intuitively, one would suspect that a Cournot equilibrium is a stationary state of this process. However, as the literature on learning in oligopoly[14] has shown, this is only one particular outcome. Moreover, a "*situation which is not a so-*

[13]Böhm & Naeve (1995) give an example of an economy where this problem arises even if there are no such expectation effects.

[14]Cf. e.g. the survey of Kirman (1995).

lution of the 'true model' but which is self-sustaining" (Kirman (1995)), or in other words an equilibrium with subjective demand, may also be a stationary state. Which equilibrium will eventually be attained, depends on the initial information and the learning process of firms, but it cannot be guaranteed that such a process converges to an equilibrium with objective demand.

In this thesis different concepts of equilibrium and dynamics in models with imperfect competition are examined. This will be done first in a static general equilibrium setting, and then in the framework of intertemporal macroeconomic models. In such models two different approaches will be contrasted, one describing sequences of temporary equilibrium with objective demand, and one in which the learning process of price–setting firms is incorporated in a dynamic model.

In detail this thesis is organized as follows:

Chapter 2 introduces a general framework which describes the interaction of agents in a dynamic model, and which allows a characterization of different equilibrium concepts as stationary states of such a dynamic process. Agents decide on their actions in each period, and they have to form expectations about the influence of their actions on payoffs. A *subjective equilibrium* is defined by a stationarity condition and by a perfect foresight condition on these expectations of agents. An *objective equilibrium* is a special case where the additional restriction on the agents' expectations is imposed that agents make no forecast errors if they deviate from their optimal decision. However, this requirement is only hypothetical and cannot be deduced from a perfect foresight condition on the expectations of agents.

In Chapter 3, this framework will be applied to general equilibrium models in which firms are imperfect competitors. This will allow a direct comparison of different equilibrium concepts in the literature with objective and subjective demand. In all these models, subjective equilibria are indeterminate due to arbitrary specifications of perceived demand curves, while objective equilibria exist only under strong assumptions on the economy. In Section 3.2 models with quantity competition are presented. The formulation is more general than usual in the literature by including the cases where firms take the prices of some "competitive goods" or the income of consumers as given. Section 3.3 considers general equilibrium models with price competition, in which firms set some prices under their control. They are distinguished between models where firms serve all the demand addressed to them, and others where firms are able to ration their consumers.

Chapter 4 considers more specific macroeconomic models with an OLG structure following the usual approach of temporary equilibrium with objective demand. The purpose of this chapter is to show that this approach is not capable of characterizing determinate intertemporal equilibria in which agents make no forecast errors. Instead, such equilibria are highly indeterminate, and there exists in particular a continuum of stationary states. In 4.2 the basic concepts of a temporary and an intertemporal equilibrium are introduced. It is essential to make a distinction between *perfect foresight equilibria* in which agents make no forecast errors along the equilibrium path and the refinement of *rational expectations equilibria* where agents also make no forecast errors off the equilibrium path. Section 4.3 shows indeterminacy of perfect foresight equilibria and the existence of involuntary unemployment in an OLG version of the model of Hart (1982). Section 4.4 shows that such indeterminacy results in general are not due to the weaker notion of perfect foresight equilibria, but that they are also valid under rational expectations. This will be illustrated in two examples.

Chapter 5 considers a macroeconomic model which is similar to that of the previous chapter, but follows a novel and strictly dynamic approach by avoiding any usage of equilibrium concepts, neither temporary nor intertemporal. Instead, in each period the labor and the goods market open sequentially, which forces firms to decide on their goods supply before they enter the goods market. Wage rates follow a competitive adjustment process, while goods prices are set by firms which believe in a demand function and try to learn its shift parameter. Wages and prices are fixed during the period implying that the labor and goods markets need not clear. Thus, this approach picks up the tâtonnement critique and the suggestions of Arrow (1959), and provides a formulation of a price adjustment process in which (at least on the goods market) price changes are the result of decisions of agents in the economy instead of an invisible hand. As already predicted by Arrow, it will be shown here that the price dynamics with more than one firm is highly unstable, displaying cyclical dynamic behavior of high orders.

The thesis concludes with a brief summary and discussion in Chapter 6.

Chapter 2

Subjective and objective equilibria

> *"(...) the significance of a reiterating process which is the central concept of the static theory in economics, can be explained only on the basis of a more general theory of economic development. If we begin with the static assumption as an isolated premiss, formulated as 'abstraction from the time factor' or something of that sort, we are easily led astray, as is shown by the laborious but not always very successful attempts 'to introduce the time factor in the Walrasian system of equations.'"*
>
> E. Lindahl (1939)

2.1 Introduction

In this chapter a basic framework will be introduced which is sufficiently general to include all models of imperfect competition to be presented in subsequent chapters. It is the intention to formulate equilibria in such models as stationary states of a certain dynamic process and to discuss features of these equilibria.

In Section 2.2 the general framework is introduced which is a simple extension of repeated normal form games. It describes interactions of agents who do not know the influence of their actions on certain payoff–relevant outcomes. As a result, agents do not know their own payoff function.

Such a starting point might not seem particularly interesting to game–theorists, but is, in my opinion, the appropriate framework for monopoly and oligopoly theory, since it is highly plausible to assume that imperfectly competitive firms have only a very crude perception of the relationship between price and demand. While the true relationship is described by an outcome function, agents in general neither know this function, nor do they perfectly foresee (and maybe even cannot observe afterwards) the actions chosen by simultaneously moving competitors. Instead, each agents has an idea of his individual outcome function describing the connection between his action and the outcome. In each period agents choose their best action given their expectations, and they use observations of past actions and outcomes to improve their expectations.

Starting from a perfect foresight condition on expectation functions, subjective equilibria are defined in Section 2.3 as stationary states of the dynamical system with perfect foresight. Objective equilibria are defined as special cases of subjective equilibria, but they can, in contrast, not be characterized by a perfect foresight condition. In Section 2.4, the dynamics with learning is introduced and it is discussed if these equilibria are the asymptotic outcome of such a learning process.

2.2 The basic framework

There is a set of agents $j \in J = \{1, \dots, n\}$, which will be in all the following models the set of imperfectly competitive firms. Time is discrete ($t = 0, 1, \dots$) and may be regarded either as real or as virtual time, i.e. it may, as in tâtonnemont processes, also describe situations where no production, trade and consumption take place until the dynamic process converges. As explained above, the methodology is to describe equilibria as stationary states of some dynamic processes.

In each period t agent j chooses an *action* s_t^j from his *action set* S^j which will describe later the choice set of production plans and/or prices. $s_t = (s_t^1, \ldots, s_t^n) \in S := \prod_{j \in J} S^j$ denotes a typical action profile at period t. Alternatively, s_t will be written as (s_t^j, s_t^{-j}), where $s_t^{-j} = (s_t^i)_{i \neq j}$. At each period there is an outcome $a_t \in A$, where A is the *outcome set.* All sets A, S^j $(j \in J)$ are subsets of some Euclidean space. It is assumed that each agent observes the outcomes, but not necessarily the actions chosen by the other agents.

The outcome in period t depends on all actions in period t, and possibly on the previous period's outcome. That is, there is an *outcome function*

$$F : A \times S \to A \quad ,$$

with the interpretation $a_t = F(a_{t-1}, s_t)$. Thus, it is assumed that at the beginning of period t all agents simultaneously choose actions $s_t = (s_t^j)_{j \in J}$ which lead to the outcome a_t. This outcome depends also on the outcome of the previous period a_{t-1}. With a sufficiently general definition of the outcome set, this formulation includes situations where the outcome function depends on past actions as well as on past outcomes more than one period before. In subsequent models of imperfect competition, the outcome function describes the reaction of the competitive sector of the economy to the actions chosen by firms, i.e. it will typically describe a direct or inverse demand function.

Each agent receives in every period a payoff described by some *individual payoff function* $\pi^j : S^j \times A \to \mathbb{R}$, such that $\pi^j(s_t^j, a_t)$ denotes the payoff of agent j in period t. Thus, the individual payoff function is assumed to depend on actions of other agents only via the outcome. Each agent is assumed to maximize the current expected payoff. The more general situation where agents maximize a discounted sum of future expected payoffs may be treated analogously, but leads to the additional difficulty that in situations with uncertainty agents may face a trade–off between maximizing the short–run payoff and gaining information by experimentation (and thereby increasing future expected payoffs). For a discussion of this problem of "active learning", see for example Kiefer & Nyarko (1989) and Mirman, Samuelson & Urbano (1993).

The collection $(J, S, A, F, (\pi^j)_{j\in J})$ will be called a *dynamic system with strategic interaction.* A complete dynamical system can be defined if additionally the expectation formation and decision making of agents are specified. It is assumed that each agent knows his individual payoff function π^j, i.e. he is aware of the direct influence of his own action and of the outcome on his payoff. However, agents do not know the outcome function F describing the relationship between the actions and the outcome. In particular they do not know their full payoff function defined by

$$\tilde{\pi}^j(s_t, a_{t-1}) := \pi^j(s_t^j, F(a_{t-1}, s_t)) \quad .$$

Moreover, agents do not foresee the actions chosen simultaneously by the other agents. As an illustration, consider the special case of a repeated normal form game:

Example 1: Consider a normal form game defined by a set of players J, strategy spaces $S = \prod_{j\in J} S^j$, and a payoff matrix $\pi : S \to \mathbb{R}^J$. The repeated play of this game can be formulated as a dynamic system with strategic interaction as follows: The outcome set is defined to be equal to the product of strategy spaces, $A = S$, and the outcome function is the identity: $F = \text{id} : S \to A$. Agents know their payoff function π^j, but they do not foresee the strategies of the other players. They have to be forecasted on the basis of past observations, as it is the case in the literature on "Learning in Games", see for example Eichberger (1992).$\diamond$

There are other ways to formulate the repeated game of the above example as a dynamic system with strategic interaction in which agents have only an incomplete knowledge of their payoff function. In the literature on evolutionary game theory (see e.g. Weibull (1995)), agents are ignorant of their payoff function, but the evolution is described by ad–hoc behavioral assumptions, while in the following the expectation and decision making process will be described explicitly.

Each agent j believes in a *subjective outcome function* $F^j : A \times S \to A$ which describes the subjective view of agent j how the outcome depends on the past outcome and the actions. Furthermore, the agent holds expectations about the choice of actions of other agents. These are described by *subjective action functions* $G^j : A \to S^{-j} := \prod_{i \neq j} S^i$. This implies in particular that agent j believes that actions of the other agents depend on the past outcome only, but not, for instance, on the own action. This is reasonable, since by assumption all actions are chosen simultaneously. Both subjective outcome and action functions together induce an *individual outcome function* (to be abbreviated IOF) by

$$\begin{aligned} \psi^j : A \times S^j &\to A \quad , \\ \psi^j(a_{t-1}, s_t^j) &:= F^j(a_{t-1}, (s_t^j, G^j(a_{t-1}))) \quad . \end{aligned}$$

Of course, if the individual payoff function of agent j does not depend on certain components of a, expectations about these components are irrelevant, but they are formally included for notational simplicity.

An IOF ψ^j is called *viable* if for all $a \in A$

$$\operatorname*{argmax}_{s^j \in S^j} \pi^j(s^j, \psi^j(a, s^j)) \neq \emptyset \quad ,$$

i.e. if the decision problem of agent j has a solution.[1] The set of viable IOFs of agent j is denoted Ψ^j. The solution correspondence is

[1] More generally, one may define IOF of agents under uncertainty by $\psi^j : A \times S^j \to \mathcal{P}(A)$, and may call them viable if $\max_{s^j \in S^j} \int_A \pi^j(s^j, a')\psi^j(a, s^j)(da')$ has a solution for all $a \in A$.

denoted $R^j(a;\psi^j) := \operatorname{argmax}_{s^j \in S^j} \pi^j(s^j, \psi^j(a, s^j))$. Denote by $r^j(a;\psi^j) \in R^j(a;\psi^j)$ selections of the solution correspondence, and take the collection $(r^j)_{j\in J}$ as given.

Given subjective outcome and action functions, and the induced IOFs $(\psi^j)_{j\in J}$, the resulting dynamical system is described by the evolution of outcomes according to

$$a_{t+1} = F\Big(a_t, (r^j(a_t;\psi^j))_{j\in J}\Big) \quad . \tag{2.1}$$

2.3 Perfect foresight and equilibrium concepts

In analogy to Böhm & Wenzelburger (1995), the following definition specifies the condition on expectation functions for which the induced dynamical system (2.1) generates trajectories with perfect foresight:

Definition 1: The collection of IOFs $(\psi^j)_{j\in J}$ is called *perfect at $a \in A$* if for all $j \in J$

$$\psi^j(a, r^j(a;\psi^j)) = F(a, (r^k(a;\psi^k))_{k\in J}) \quad .$$

$(\psi^j)_{j\in J}$ is called *perfect* if it is perfect at a for all $a \in A$. Solutions of (2.1) with perfect IOFs are *perfect foresight trajectories.*

Notice that perfect foresight of actions of the other agents is not explicitly required in this definition since it is assumed that agents only observe outcomes. If agents also observe actions of the other agents, they could be formally included in this formulation by extending the outcome space to $\hat{A} := A \times S$ and by extending the outcome function by the identity map in the second component to $\hat{F} : \hat{A} \times S \to \hat{A}$. If one assumes additionally that the subjective outcome functions $F^j : \hat{A} \times S \to \hat{A}$ coincide with the identity map in the second component, then perfect foresight of actions of the other agents

$$(r^k(a;\psi^k))_{k\in J,k\neq j} = G^j(a)$$

follows from the above condition.

It should be emphasized that in this framework perfect foresight cannot be expressed by independent individual conditions on expectation functions. Instead, the perfect foresight condition of each agent depends on the expectation functions of the other agents. Thus, perfect foresight conditions of all agents are mutually dependent.

It is natural to define an equilibrium as a steady state of the dynamical system (2.1) with perfect outcome expectations. More precisely:

Definition 2: $(\bar{s},\bar{a})$ is a *(subjective) equilibrium* of $(J,S,A,F,(\pi^j)_{j\in J})$ if there are IOFs $(\psi^j)_{j\in J}$ such that

(i) $\bar{a} = F(\bar{a},\bar{s})$,

(ii) $\bar{s}^j = r^j(\bar{a};\psi^j) \quad \forall\, j\in J$,

(iii) $(\psi^j)_{j\in J}$ is perfect at $\bar{a}$, i.e. $\psi^j(\bar{a},\bar{s}^j) = \bar{a}$.

Even though perfect foresight is already a strong requirement on the expectations of agents, there still remains a large degree of arbitrariness of perfect foresight trajectories in these definitions. The explanation of this phenomenon is the fact that agents form expectations about *outcome functions*, but observe only those *outcomes* where all agents decide optimally contingent upon their expectations. Thus, individual outcome *functions* cannot be determined by a perfect foresight condition, and this arbitrariness may lead to different trajectories with perfect foresight and different equilibria. Such multiplicities are not due to the interaction of different agents, as already the textbook example of a monopolist shows:

Example 2: Consider a quantity setting monopolist supplying a homogeneous good with strictly decreasing inverse demand function $F:\mathbb{R}_+\to$

$\mathbb{R}_+$, $p = F(q)$, and constant marginal costs c. That is, $q \in S := \mathbb{R}_+$, $p \in A := R_+$, and $\pi : S \times A \ni (q,p) \mapsto (p-c)q \in \mathbb{R}$. The outcome function F does not depend on past outcomes. Consider an arbitrary $\tilde{p} \in [c, F(0))$. Then any linear IOF $\psi_{\tilde{p}}(q) := \tilde{a} - \tilde{b}q$ with $\tilde{a} := 2\tilde{p} - c$ and $\tilde{b} := (\tilde{p} - c)/F^{-1}(\tilde{p})$ is perfect since $\tilde{q} := F^{-1}(\tilde{p})$ maximizes the expected profit $(\psi_{\tilde{p}}(q) - c)q$ and $\psi_{\tilde{p}}(\tilde{q}) = \tilde{p} = F(\tilde{q})$.

Moreover, consider an arbitrary map $\Phi : [c, F(0)) \to [c, F(0))$. Then the trajectories $(p_t)_{t\geq 0}$ generated by this map (i.e. $p_{t+1} = \Phi(p_t)$) coincide with those of (2.1) for the perfect IOF $\psi : A \times S \to A$ defined by $\psi(p_-, q) := \psi_{\Phi(p_-)}(q)$. Given p_{t-1}, $q_t = F^{-1}(\Phi(p_{t-1}))$ maximizes the expected profit in period t $(\psi(p_{t-1}, q) - c)q$, and therefore

$$F(q_t) = F(F^{-1}(\Phi(p_{t-1}))) = \Phi(p_{t-1}) = p_t \ .$$

Thus, even though the outcome function depends only on the current action but not on past outcomes, there are perfect IOFs depending on the past outcome and leading to arbitrary trajectories with perfect foresight! Moreover, each $(F^{-1}(\tilde{p}), \tilde{p})$ with $\tilde{p} \in [c, F(0))$ is an equilibrium according to Definition 2. ◇

As the above example shows, even in the simplest case of a single–good monopolist there is a continuum of equilibria. However, almost all of them are *non–objective* equilibria, i.e. they do not maximize the true payoff function $(F(q)-c)q$. Therefore, most of the literature on imperfect competition restricts attention to objective equilibria:

Definition 3: An IOF ψ^j is *objective at* (a, s^{-j}) if

$$\psi^j(a, s^j) = F(a, (s^j, s^{-j})) \qquad \forall s^j \in S^j \ .$$

An equilibrium $(\bar{s}, \bar{a})$ with corresponding IOFs $(\psi^j)_{j\in J}$ is *objective* if ψ^j is objective at $(\bar{a}, \bar{s}^{-j})$ for all $j \in J$.

Objective IOFs coincide with the outcome function not only in the equilibrium point, but even outside, provided that the actions of the other agents do not deviate from their equilibrium values. In particular, objectiveness is again defined conditional on actions of other agents and thereby on their IOFs. However, the definition differs conceptually from the perfect foresight condition in Definition 1 since it ties down expectations of agents in situations which actually never happen as long as agents maximize payoffs. Hart (1985) remarks that "*what distinguishes the objective demand approach from the subjective approach is the explicit modeling of a non–cooperative game that firms play.*" However, a game is modeled in the subjective approach as well. What makes the difference is an assumption on outcome expectations of agents.

It is a straightforward implication of this definition that trajectories $(a_t)_{t\geq 0}$ of (2.1) with IOFs which are objective in each period coincide with sequences of Nash equilibria of the normal form games with payoff functions $\tilde{\pi}^j_t(s) := \pi^j(s^j, F(a_t, s))$. Existence of a Nash equilibrium typically requires strong assumptions. The usual sufficient existence condition on the quasi–concavity of the resulting payoff functions $\tilde{\pi}$ corresponds to conditions on the outcome function F, which may not be satisfied in some of the models to be presented in subsequent chapters.

It is reasonable to think of agents using finitely parameterized models. If the outcome function F is very complicated, it is very implausible that agents actually believe in their true objective outcome function. A reply to this objection is the argument that in many situations not the whole objective function has to be known, but that it suffices to know the derivative of it. This leads to the concept of locally objective equilibria:[2]

Definition 4: Let F be differentiable in s. An IOF ψ^j is *locally objective*

[2]Cf. Silvestre (1977) and Bonanno & Zeemann (1985).

at (a, s) if it is differentiable in s^j and if[3]

$$d_{s^j} F(a, (s^j, s^{-j})) = d_{s^j} \psi^j(a, s^j) \quad .$$

An equilibrium $(\bar{s}, \bar{a})$ with corresponding IOFs $(\psi^j)_{j \in J}$ is *locally objective* if ψ^j is locally objective at $(\bar{a}, \bar{s})$ for all $j \in J$.

The following result is a straightforward extension of a theorem proven in Gary-Bobo (1989) for Cournot–Walras equilibria:

Theorem 1: Assume that for all $j \in J$ the set S^j is convex and that for all (a, s^{-j}) the function $s^j \mapsto \pi^j(s^j, F(a, (s^j, s^{-j})))$ is strictly quasi–concave, C^2 and that all critical points are non–degenerate (i.e. the second derivative at a critical point is non–singular). Then every locally objective equilibrium is an objective equilibrium.

Proof: Let $(\bar{s}, \bar{a})$ be a locally objective equilibrium. From $\bar{s}^j \in R^j(\bar{a}; \psi^j)$ and the convexity of S^j it follows

$$\Big(d_{s^j}\pi^j(\bar{s}^j, \bar{a}) + d_a\pi^j(\bar{s}^j, \bar{a}) d_{s^j}\psi^j(\bar{a}, \bar{s}^j)\Big)(\bar{s}^j - s^j) \geq 0 \quad \forall s^j \in S^j.$$

Defining $\tilde{\pi}^j(s^j) := \pi^j(s^j, F(\bar{a}, (s^j, \bar{s}^j)))$, local objectivity implies

$$d\tilde{\pi}^j(\bar{s}^j)(\bar{s}^j - s^j) \geq 0 \quad \forall s^j \in S^j \ . \tag{2.2}$$

Assume first that $d\tilde{\pi}^j(\bar{s}^j) = 0$. Since by assumption $\tilde{\pi}^j$ is non–degenerate and strictly quasi–concave, $d^2\tilde{\pi}^j(\bar{s}^j)$ is negative definite, and thus $\bar{s}^j$ is a local maximizer. Since $\tilde{\pi}^j$ is strictly quasi–concave on the convex set S^j, $\bar{s}^j$ is a global maximizer. Assume second that $d\tilde{\pi}^j(\bar{s}^j) \neq 0$ and suppose that $\bar{s}^j$ is not a global maximizer. This implies that there exists $s^j \in S^j$ such that $\tilde{\pi}^j(s^j) > \tilde{\pi}^j(\bar{s}^j)$. But now strict quasi–concavity of $\tilde{\pi}^j$ and convexity of S^j imply

$$d\tilde{\pi}^j(\bar{s}^j)(\bar{s}^j - s^j) < 0 \ ,$$

[3] $d_x f(x, y)$ denotes the differential of f at (x, y) with respect to the argument x.

a contradiction to (2.2). Therefore, in any case $\bar{s}^j$ is a global maximizer of $\tilde{\pi}^j$, which implies that $(\bar{s}, \bar{a})$ is an objective equilibrium. □

Notice that the assumptions of Theorem 1 together with compactness of the sets A and S imply the existence of an objective equilibrium as a fixed point of the continuous mapping

$$(s, a) \mapsto \left(\left(\operatorname*{argmax}_{\tilde{s}^j} \pi^j(\tilde{s}^j, F(a, (\tilde{s}^j, s^{-j}))) \right)_{j \in J}, F(a, s) \right) .$$

Therefore under conditions which are usually assumed in order to get existence of an objective equilibrium, it suffices to find a locally objective equilibrium. However, in situations where payoff functions fail to be quasi–concave, actions in a locally objective equilibrium need not be global maximizers of the objective payoff functions and can be even local minimizers.

2.4 Learning dynamics

The perfect foresight condition has been imposed in the previous section in order to describe situations where agents make no forecast errors. While this may be regarded as a desirable situation, it is not a very realistic description of the world. Agents should be modeled more plausibly to have certain perceptions about outcomes and the role of their actions on outcomes, and to revise these perceptions if they make forecast errors. Thus, the formation of expectations depends on some initial perception of the world and on current and past observations of agents. Such kinds of processes may generally be called *learning dynamics.*

According to Marimon (1997) the most important arguments in favor of learning in economics are *bounded rationality*, *equilibrium justification* and *equilibrium selection.* While the first aspect corresponds to the argument described above, the other two aspects play also an important role in this context: Can the equilibria defined in 2.3 not only be attained as a result of a fixed point argument, but also as the asymptotic outcome of a (more or less reasonable) learning dynamics? Does learning dynamics select equilibria of the large set of subjective equilibria, and in particular, can objective equilibria be selected by learning dynamics?

When agents form expectations about outcomes and actions, i.e. when they choose subjective outcome and action functions, they only know a history of past outcomes and past actions. The set of histories at the beginning of period t is denoted

$$H_{t-1} := A \times \prod_{\tau=1}^{t-1} (S \times A) \quad ,$$

and a typical element is $h_{t-1} = (a_0, (s_\tau, a_\tau)_{\tau=1}^{t-1})$.

Definition 5: A *learning process* of agent j is a sequence of mappings

$$L^j = (L_t^j)_{t=1}^{\infty} \quad , \quad L_t^j : H_{t-1} \to \Psi^j .$$

Thus, in each period agent j chooses a viable IOF $\psi_t^j = L_t^j(h_{t-1})$. For notational simplicity, there is no explicit distinction as in 2.2 between the formation of outcome and action expectations.

Given the dynamic system with strategic interaction $(J, S, A, F, (\pi^j)_{j\in J})$ and learning processes L^j, the resulting dynamical system with learning is described by

$$s_t = \left(r^j(a_{t-1}; L_t^j(h_{t-1})) \right)_{j\in J} \quad , \tag{2.3}$$

$$a_t = F(a_{t-1}, s_t) \quad , \tag{2.4}$$

$$h_t = (h_{t-1}, (s_t, a_t)) \quad . \tag{2.5}$$

According to the above definition, learning processes are defined very generally and allow for a large variety of possible learning dynamics and resulting equilibria. However, one would not accept situations as limit outcomes of the learning dynamics, where agents make permanent and systematic forecast errors. In order to exclude such situations, one has to impose consistency requirements on the learning process. In particular, it will be required here that if observations converge, then the IOFs selected by the learning process have to converge to an IOF with perfect foresight.[4]

Definition 6: The learning process L^j is called *consistent*, if for all sequences (s_t, a_t, h_t) fulfilling (2.5) with $\lim_{t\to\infty} s_t = \bar{s}$ and $\lim_{t\to\infty} a_t = \bar{a}$, there exists a continuous $\psi^j \in \Psi^j$ such that $\psi_t^j := L_t^j(h_{t-1})$ converges locally uniformly to ψ^j and $\psi^j(\bar{a}, \bar{s}^j) = \bar{a}$.

It follows from this definition that forecast errors converge to zero if actions and outcomes converge. Indeed, continuity of ψ^j and locally uniform convergence of ψ_t^j to ψ^j imply:

$$\begin{aligned} |\psi_t^j(a_{t-1}, s_t^j) - a_t| &\leq |\psi_t^j(a_{t-1}, s_t^j) - \psi^j(a_{t-1}, s_t^j)| + \\ &\quad |\psi^j(a_{t-1}, s_t^j) - \psi^j(\bar{a}, \bar{s}^j)| + |\bar{a} - a_t| \stackrel{t\to\infty}{\longrightarrow} 0 \quad . \end{aligned}$$

It is an immediate consequence of this consistency requirement that the set of steady states of (2.3)-(2.5) for consistent learning processes coincides with the set of subjective equilibria, i.e. subjective equilibria are completely characterized (or "justified") by consistent learning processes:

Theorem 2: Assume that F and π^j are continuous and let $(\bar{s}, \bar{a}) \in S \times A$. Then

(i) If $(s_t, a_t, h_t)_{t\geq 1}$ is a trajectory of (2.3)-(2.5) with a consistent learning process $(L^j)_{j\in J}$ such that $(s_t, a_t) \to (\bar{s}, \bar{a})$, then $(\bar{s}, \bar{a})$ is a subjective equilibrium.

[4]In non–converging situations, other consistency requirements should be imposed on the learning process. For instance, one may claim that autocorrelations of forecast errors converge to zero (cf. Hommes & Sorger (1998) and Schönhofer (1997)).

(ii) If $(\bar{s}, \bar{a})$ is a subjective equilibrium, then there is a consistent learning process $(L^j)_{j\in J}$ such that $(\bar{s}, \bar{a}, \bar{h}_t)_{t\geq 1}$, where $\bar{h}_t$ is defined by $\bar{h}_0 = \bar{a}$ and $\bar{h}_t = (\bar{h}_{t-1}, (\bar{s}, \bar{a}))$, is a trajectory of (2.3)-(2.5).

Proof:

(i) By continuity of F and consistency it follows $F(\bar{a}, \bar{s}) = \bar{a} = \psi^j(\bar{a}, \bar{s}^j)$. It remains to show that $\bar{s}^j \in R^j(\bar{a}; \psi^j)$. Suppose this is not the case, i.e. there exists s^j such that $\pi^j(s^j, \psi^j(\bar{a}, s^j)) > \pi^j(\bar{s}^j, \psi^j(\bar{a}, \bar{s}^j))$. But then $s^j_t \in R^j(a_{t-1}; \psi^j_t)$ for all $t \geq 1$ implies

$$\begin{aligned}\pi^j(\bar{s}^j, \psi^j(\bar{a}, \bar{s}^j)) &= \lim_{t\to\infty} \pi^j(s^j_t, \psi^j_t(a_{t-1}, s^j_t)) \\ &\geq \lim_{t\to\infty} \pi^j(s^j, \psi^j_t(a_{t-1}, s^j)) = \pi^j(s^j, \psi^j(\bar{a}, s^j)) \quad ,\end{aligned}$$

a contradiction.

(ii) Take the IOFs $(\psi^j)_{j\in J}$ of Definition 2. Define L^j_t by $L^j_t(\bar{h}_{t-1}) = \psi^j$ and arbitrary and consistent if $h_{t-1} \neq \bar{h}_{t-1}$ (for instance, by adaptive outcome expectations $L^j_t(h_{t-1})(a_{t-1}, s^j) = a^e_{t-1} + \lambda(a_{t-1} - a^e_{t-1})$). Then clearly $(\bar{s}, \bar{a}, \bar{h}_t)_{t\geq 1}$ is a trajectory of (2.3)-(2.5). □

As an illustration of this result, consider a monopolist who believes in a linear inverse demand curve with intercept a and negative slope $-b$. A consistent learning process may be defined by fixing one parameter a or b and updating of the other parameter. The resulting learning dynamics would have different steady states depending on the fixed parameter. One can think as well of learning processes with updating of both parameters and which have a continuum of steady states. This is the case in models where the monopolist estimates the parameters with an OLS method, cf. Rampa (1989).

It is a consequence of the above result that further assumptions on the learning process have to be imposed in order to select among the set of subjective equilibria by learning dynamics. An agent's learning process reflects his subjective perception of the economy and cannot be fixed "objectively" by an outside observer. All one can do is to exclude too stupid agents by imposing consistency or rationality requirements on the learning process. It is, however, not clear if there exist learning processes which select (locally) objective equilibria *independently* of the true outcome function. This will be discussed briefly in the following.

A first observation is that if it were assumed that agents not only observe past histories of outcomes and actions, but also the derivatives of the outcome function at past states, a corresponding consistency requirement on the learning process could claim that forecast errors of the derivatives vanish asymptotically. The only stationary limits of learning processes fulfilling this requirement are then be locally objective equilibria. Such an assumption might be justified by the idea that agents carry out small, but fictitious experiments in every period from which they can estimate the derivative. Since such experiments are not modeled explicitly, it will be assumed that agents do not have this additional information.

Since in a locally objective equilibrium the action–derivative of the IOF has to coincide with the action–derivative of the outcome function, it is a necessary condition that the learning process is sufficiently parameterized in order to be capable of approximating locally objective IOFs. For instance, in the example of a monopolist with linear demand perception, both parameters a and b have to be updated, since updating of only one parameter yields only incidentally the objective equilibrium as a steady state.

However, even if the monopolist has a model with two free parameters, there may still remain a continuum of steady states. In order to select a locally objective equilibrium, one may think of a learning process in which the monopolist tries to estimate the slope of the unknown inverse demand function from recent past observations. Suppose that the monopolist believes in a linear inverse demand $p = a - bq$ and uses the following updating scheme:

$$b_{t+1} = \begin{cases} \dfrac{p_t - p_{t-1}}{q_{t-1} - q_t} & \text{if } q_t \neq q_{t-1} , \\ b_t & \text{else} , \end{cases} \qquad a_{t+1} = p_t + b_{t+1} q_t \quad ,$$

where $q_t = \text{argmax}(a_t - b_t q - c)q$ and $p_t = F(q_t)$. With this learning process, for all converging, but non–constant trajectories the slopes of the IOFs necessarily converge to the true slope of the inverse demand function at the equilibrium, since

$$\lim_{t\to\infty} \frac{F(q_t) - F(q_{t-1})}{q_{t-1} - q_t} = -F'(\bar{q}) \quad \text{if} \quad \lim_{t\to\infty} q_t = \bar{q} \ .$$

Thus, any non–constant converging sequence converges to a locally objective equilibrium. Other equilibria are still steady states, but either trajectories jump directly on it or they do not converge to them.

The situation changes considerably in a multi–product monopoly or in an oligopoly. If a trajectory converges to a state, typically only a directional derivative of the outcome function may be estimated from past observations. Thus, in general it seems impossible to obtain enough information from observations such that the local objectivity condition of Definition 4 is necessarily satisfied asymptotically.

An argument in favor of objective equilibria is that they are "stable" under unilateral experimentation. That is, if exactly one agent changes his action, he will observe what he expects (for any change in an objective equilibrium or for small changes approximately in a locally objective equilibrium). However, this intuitive stability argument is just informal since experimentation is not modeled explicitly. Agents only have incentives to experiment, if they maximize intertemporally and if they are uncertain about their model. Both extensions of the framework have been excluded so far, since, as Kirman (1995) remarks, "*to incorporate this problem of 'active learning' into a model in which there is also strategic interaction is, for the moment, unmanageable.*" But even in the case of a monopolist facing an unknown linear demand curve and trying to learn the parameters with Bayesian updating, convergence against a (not necessarily objective!) equilibrium can be shown only if the true parameter values lie in the support of the prior distribution (i.e. if the model is correctly specified), see Kiefer & Nyarko (1989). If the model is not correctly specified, then also cycles may occur (Nyarko (1991)).

2.5 Summary

The concepts of subjective and objective equilibria in models with strategic interaction but with ignorance of payoff–relevant outcome functions have been introduced in a general framework. Subjective equilibria are typically completely indeterminate, but may be characterized by a perfect foresight condition and also as stationary outcomes of consistent learning processes. Objective equilibria are special cases, but there are no indications why objective equilibria should be more likely limiting outcomes of an (even very sophisticated) learning process.

Nevertheless, objective equilibria are selections of subjective equilibria with respect to stability under unilateral, although fictitious experimentation, and they have been intensively investigated in the literature. In the following chapters, some models of imperfect competition will be presented for which objective equilibria and, in some earlier work, also subjective equilibria have been defined.

Chapter 3

General equilibrium with imperfect competition

"The traditional 'market demand curve' for a certain product is not the same sort of thing as the demand curve which is relevant in determining the actions of the individual producer. The first denotes a functional relationship between the price and the amounts bought from a particular producer. The second concerns the image of this functional relationship as it exists in the mind of the entrepreneur. The two may differ widely."

N. Kaldor (1934)

3.1 Introduction

It is the purpose of this chapter to present several models of general equilibrium with imperfect competition and to show how they fit into the framework of Chapter 2. All of these models are static in the sense that all trade, consumption and production decisions take place at one stage, i.e. time has no meaning in these models. Nevertheless, the purpose is to give the various equilibrium concepts a dynamic interpretation, as it is possible to understand a perfectly competitive equilibrium as stationary solution of tâtonnement processes. In a similar spirit, in this chapter specific dynamic systems with strategic interactions and a fictitious time structure will be specified in order to provide (pseudo–)dynamic foundations of different equilibrium concepts with imperfect competition.

First of all, two fundamental questions arise: Why does a general equilibrium theory need imperfectly competitive agents, and why does a theory of imperfect competition need a general equilibrium formulation?

The first question can be answered from the methodological position that a useful economic theory should be based on a more realistic assumption than that of price–taking behavior of all agents. In real world economies there are agents (firms, unions, public authorities) who are non–price–takers, i.e. who believe that their behavior influences prices or who have direct administrative price setting power. A further motivation for considering imperfect competition is the fact that general equilibrium theory lacks a description of price formation. That is, if all agents are price–takers, " *there is no one left over whose job it is to make a decision on price* " (Arrow (1959)). In particular, when applying the methodology of microeconomic foundation consistently, one would " *want price changes to be the outcome of the calculations of rational optimizing agents and not hand them over to an auctioneer* " (Hahn (1987)). Thus, imperfectly competitive agents are proposed to take over the part of price formation in a general equilibrium model.[1]

There is a huge literature on *industrial organization* dealing with the theory of monopoly, oligopoly and monopolistic competition in a partial market equilibrium framework. An answer to the second question is that

[1]Of course, the Cournotian approach does not answer this (second) purpose, if it is not based on a two–stage quantity/price competition as in Kreps & Scheinkman (1983). To my knowledge, however, this has never been done in a general equilibrium model.

in this framework interactions with factor markets (i.e. factor utilization and feasibility constraints) and income effects (i.e. the influence of factor incomes on demand) are widely ignored. These aspects are, however, particularly important in a macroeconomic model.[2]

At least it is possible to integrate the standard textbook partial equilibrium models in a "general" equilibrium framework by use of quasi-linear preferences such that wealth effects do not matter (cf. Mas-Colell, Whinston & Green (1995), Chapters 10 and 12). Such a partial market modeling is, because of its simplicity, the appropriate framework for many important questions in industrial organization not under consideration here, such as entry deterrence, endogenous product differentiation, and others. A general equilibrium modeling is intended to complete this theory and to check its consistency.

It should be noticed, however, that the problems concerning existence of equilibria in the models to be presented below, are not typical of general equilibrium analysis. They may as well appear in partial equilibrium models where strong assumptions on exogenous demand curves have to be imposed in order to get existence of equilibria.

This chapter is organized as follows: In Section 3.2 models with quantity competition will be presented. The starting point is the assumption that a firm's action is a production plan, and that the outcome function determines goods' prices. Cournot–Walras equilibria as defined by Gabszewicz & Vial (1972) are objective equilibria where firms take into account the reactions of all prices to their production decision. Other objective equilibria are also defined, where firms take prices of some "competitive goods" and/or the profit incomes of consumers as given. The subjective equilibrium concept defined by Negishi (1961, 1972) turns out to be a special case.

Section 3.3 describes models where the formation of prices is handed over

[2]Some authors (e.g. Mankiw (1985)) still discuss partial equilibrium models and give them a macroeconomic interpretation.

to firms. They may be distinguished between models in which firms serve the Walrasian demand of the competitive sector, even if they do it non-voluntarily, and models in which firms may ration consumers and where all spillovers between different markets are taken into account. Objective equilibria for the first type of models have been defined by Marshak and Selten (1974) and Hart (1985), while subjective and objective equilibria for the second type of models have been defined by Benassy (1976 and 1988).

3.2 Quantity competition

3.2.1 The setup

Consider a private ownership economy

$$\left((X_i, \succsim_i, \omega_i)_{i\in I}, (Y_j)_{j\in J}, (\theta_{ij})_{i\in I, j\in J}\right)$$

with the usual interpretation: $I = \{1, \ldots, m\}$ and $J = \{1, \ldots, n\}$ are the sets of consumers and firms with typical elements i and j respectively. There are ℓ commodities, $l \in L = \{1, \ldots, \ell\}$. $X_i \subset \mathbb{R}^\ell, i \in I$, and $Y_j \subset \mathbb{R}^\ell, j \in J$, are the consumption and production sets. $\succsim_i$ are preference relations on X_i to be represented by utility functions $u_i : X_i \to \mathbb{R}$. $\omega_i \in \mathbb{R}^\ell$ is the initial endowment of consumer i, and θ_{ij} his share of the profits of firm j.

It will be assumed that $X_i = \mathbb{R}^\ell_+$ and that preferences are continuous, strictly convex, and strictly monotone, such that the solution of consumer i's utility maximization problem is described by the Walrasian demand function

$$x_i : \mathbb{R}^\ell_{++} \times \mathbb{R}_+ \ni (p, w) \mapsto x_i(p, w) \in \mathbb{R}^\ell_+$$

which fulfills $p \cdot x_i(p, w) = w$ for all price/income–pairs (p, w).

Firms are strategic agents, and in this section it is assumed that they are quantity setters, i.e. their action is a choice of a production plan.

Therefore, $S^j = Y_j$ and $s^j = y_j \in Y_j$. As usual for Cournot competition, after all firms choose production plans $y = (y_j)_{j \in J}$, an inverse demand function determines some prices while there may be other prices that remain unaffected. More precisely, assume that there are ℓ^s *strategic* goods $L^s \subset L$ whose prices $p^s \in \mathbb{R}^{\ell^s}_{++}$ adjust in every period immediately (i.e. infinitely *fast*) to the market clearing level. The remaining $\ell^c = \ell - \ell^s$ *competitive* goods $L^c = L \setminus L^s$ have prices $p^c \in \mathbb{R}^{\ell^c}_{++}$ adjusting *slowly*, i.e. from period to period. They are announced at the beginning of each period by an auctioneer who adjusts them according to the excess demand of the previous period. Thus, production decisions of firms in the actual period have no effect on prices of competitive goods, but they can have an effect on prices of strategic goods.[3]

Profit incomes of consumers may, as above, also adjust fast or slowly. First the case of fast income adjustment will be treated. Outcomes are price/production pairs $a = (p, y) \in A := \mathbb{R}^{\ell}_{++} \times Y$ where $Y := \prod_{j \in J} Y_j$. An outcome function is described by

$$\begin{aligned} F : A \times S &\rightarrow A , \\ \Big((\tilde{p}, \tilde{y}), y\Big) &\mapsto (p, y) . \end{aligned}$$

$(\tilde{p}, \tilde{y})$ and (p, y) are the price/production–vectors of the previous and actual period.

If $y \in Y$ is a vector of production plans, denote by

$$\omega_i(y) := \omega_i + \sum_{j \in J} \theta_{ij} y_j$$

the modified endowment vector or real income of consumer i. Firms distribute outputs and collect factor inputs from consumers corresponding to their profit shares. This implies that for any vector of commodity prices the values of the distributed shares of production vectors equal the

[3] The terminology of competitive and strategic goods has been adopted from Dierker & Grodal (1995), although not with this dynamic interpretation.

profit incomes of consumers. In the contribution of Gabszewicz & Vial (1972) factor markets are ignored by the assumption $Y_j \subset \mathbb{R}^{\ell}_{+}$, with the interpretation that consumers provide firms with factors of production which are not marketable and are thus not modeled. Instead, consumers hold output shares and firms distribute the real factor income to the consumers in terms of these preassigned shares.

The definition of the outcome function is as follows: First, prices of competitive goods $l \in L^c$ follow a tâtonnement dynamics

$$p_l = G_l(\tilde{p}_l, z_l(\tilde{p}, \tilde{y})) \quad , \tag{3.1}$$

where $z_l(\tilde{p}, \tilde{y}) := \sum_{i \in I} \left(x_{il}(\tilde{p}, \tilde{p} \cdot \omega_i(\tilde{y})) - \omega_{il}(\tilde{y}) \right)$ is excess demand on market l in the previous period. The functions $G_l : \mathbb{R}_{++} \times \mathbb{R} \to \mathbb{R}_{++}$ satisfy the following

Assumption:

For at least $\ell^c - 1$ goods $l \in L^c$, G_l is strictly increasing in z_l and fulfills $G_l(p, 0) = p$ for all p.

The tâtonnement dynamics (3.1) determines the vector of competitive prices $p^c = (p_l)_{l \in L^c}$, irrespective of the current production decisions of firms.

Prices of strategic goods are assumed to adjust immediately to their market clearing level. Formally, define the inverse demand correspondence

$$\mathcal{W}^s(y, p^c) := \left\{ p^s \in \mathbb{R}^{\ell^s}_{++} \mid p := (p^s, p^c) \text{ fulfills } z_l(p, y) = 0 \ \forall\, l \in L^s \right\} \quad ,$$

and assume that $\mathcal{W}^s(y, p^c) \neq \emptyset$ for all $(y, p^c) \in Y \times \mathbb{R}^{\ell^c}_{++}$.[4] Choose any selection $P^s : Y \times \mathbb{R}^{\ell^c}_{++} \to \mathbb{R}^{\ell^s}_{++}$, $P^s(y, p^c) \in \mathcal{W}^s(y, p^c)$, so that strategic prices are defined by

$$p^s = P^s(y, p^c) \quad .$$

[4]This is a strong assumption in this generality. In subsequent sections on Cournot–Walras equilibria, it will be examined in more detail.

If $L^c = \emptyset$, the selection P^s requires a choice of price normalization since all z_l are zero–homogeneous in p, see Section 3.2.3 below. If $L^c \neq \emptyset$, prices are normalized by the choice of the tâtonnement process. For instance, some good $l_0 \in L^c$ is a numeraire if $G_{l_0}(p, z) \equiv 1$.

Consider next the case of slow profit income adjustment, where consumers' profit incomes are assumed to be determined in the previous period and thereby independent of current production decisions.[5] The outcome set and outcome function are now defined as follows: Outcomes consist of a price vector p, a collection of production plans y, and a non–negative vector of consumers' profit incomes $\pi = (\pi_i)_{i \in I} \in \mathbb{R}^m_+$. Thus, $A := \mathbb{R}^\ell_{++} \times Y \times \mathbb{R}^m_+$. The outcome function may be written

$$F : A \times S \ni \Big((\tilde{p}, \tilde{y}, \tilde{\pi}), y\Big) \mapsto (p, y, \pi) \in A \ ,$$

where $(\tilde{p}, \tilde{y}, \tilde{\pi})$ and (p, y, π) are the price/production/profit–vectors of the previous and actual period. This outcome function is defined as follows: Excess demand on market l now depends also on the predetermined profit income and is defined by

$$\hat{z}_l(p, y, \pi) := \sum_{i \in I} \Big(x_{il}(p, p \cdot \omega_i + \pi_i) - \omega_{il}\Big) - \sum_{j \in J} y_{jl} \quad .$$

Competitive prices adjust as before according to

$$p_l = G_l(\tilde{p}_l, \hat{z}_l(\tilde{p}, \tilde{y}, \tilde{\pi})) \quad , \ l \in L^c,$$

which determines the vector of competitive prices p^c. Profit incomes are determined by the previous period's dividend payments:

$$\pi_i := \sum_{j \in J} \theta_{ij} \tilde{p} \cdot \tilde{y}_j \quad .$$

[5]An alternative formulation resulting in the same set of objective equilibria would be to define a large number of replications of the above economy, to be interpreted as different but identical "islands" (cf. Hart (1985)). Consumers receive profit income on all islands, but trade only in their home island. If the number of islands is very large, the production decision of a single firm has only a negligible influence on consumers' profit income.

Finally, strategic prices clear markets of strategic commodities. Assume that for all $(y, p^c, \pi) \in Y \times \mathbb{R}^{\ell c}_{++} \times \mathbb{R}^m_+$

$$\mathcal{W}^s(y, p^c, \pi) := \left\{ p^s \in \mathbb{R}^{\ell s}_{++} \mid \hat{z}_l((p^s, p^c), y, \pi) = 0 \; \forall \, l \in L^s \right\} \neq \emptyset ,$$

and denote by $P^s(y, p^c, \pi) \in \mathbb{R}^{\ell s}_{++}$ a selection. Thus, strategic prices are determined by

$$p^s = P^s(y, p^c, \pi) \quad .$$

Finally, it will be assumed that firms maximize profits, so that payoff functions $\pi^j : S^j \times A \to \mathbb{R}$ are $\pi^j(y_j, (p, y)) := p \cdot y_j$ or $\pi^j(y_j, (p, y, \pi)) := p \cdot y_j$ respectively.

3.2.2 Negishi equilibria

Equilibria defined by Negishi (1961, 1972) turn out to be subjective equilibria of the above defined dynamic system with strategic interaction under certain restrictions on IOFs $\psi^j : A \times S_j \to A$. Since only the price–component of the outcome is payoff–relevant, it will be assumed that firms form only price (or inverse demand) expectations. Furthermore, since only a stationary equilibrium will be described, the dependence of ψ^j on the past outcome will also be suppressed. Therefore, the IOFs may be described w.l.o.g. by subjective inverse demand functions $P_j : Y_j \to \mathbb{R}^{\ell}_{++}$. Denote by P_{jl} the subjective inverse demand of firm j of good l. Negishi restricts these subjective inverse demand functions to satisfy the following

Assumption:

There is a disjoint partition of the set of goods $L = \bigcup_{j \in J} L_j \cup L_o$ such that for all j [6]

[6] In his 1962 paper, Negishi actually assumes in (2) that P_{jl} only depends on y_{jl}. Moreover, Negishi assumes that P_{jl} are affine linear functions. Hart (1985) weakens these assumptions.

1. If $l \notin L_j$, P_{jl} is constant and

2. if $l \in L_j$, P_{jl} depends only on $(y_{jl'})_{l' \in L_j}$ and is strictly decreasing in y_{jl}.

Thus, L_j is the set of goods for which firm j but no other firm believes to have an influence on their prices, and these prices are believed to depend only on firm j's supply and demand of its subjective strategic goods L_j.

Definition 1: (y^*, p^*) is a *Negishi equilibrium*, if there are $(P_j)_{j \in J}$ satisfying the above assumption such that

$$
\begin{array}{ll}
\text{(i)} & P_j(y_j^*) = p^* \;\text{ and }\; y_j^* \in \operatorname{argmax}_{y_j \in Y_j} P_j(y_j) \cdot y_j \quad \forall\, j \in J, \\
\text{(ii)} & z_l(p^*, y^*) = 0 \quad \forall\, l \in L\,.
\end{array}
$$

It is an immediate consequence of this definition that the set of Negishi equilibria coincides with the set of subjective equilibria of every dynamic system with strategic interaction specified in the previous section, but with the above restriction on subjective outcome functions. Condition (i) is the same as $\psi^j(\bar{a}, \bar{s}^j) = \bar{a}$ and $\bar{s}^j \in R^j(\bar{a}; \psi^j)$. By definition of the above outcome functions, the market clearing condition (ii) is automatically satisfied for strategic goods, and coincides with the stationarity condition of prices of competitive goods, possibly together with stationarity of profit incomes.

As a consequence, no matter how the economy actually evolves, i.e. whether prices and incomes adjust fast or slowly (or if prices and incomes are determined in some large market), Negishi equilibria only depend on subjective inverse demand functions. Since it is completely arbitrary how the set of goods is partitioned and since there are no restrictions on subjective inverse demand functions, Negishi equilibria are highly indeterminate. Gary-Bobo (1989) has shown that in fact every arbitrary production plan with non–negative profits at the equilibrium price vector is a Negishi equilibrium, albeit with a less restrictive condition on subjective inverse demand functions.

Furthermore, the assumption imposed on subjective inverse demand functions that different firms have to believe to be able to influence different prices is – unless for very exceptional model specifications – never objective. Hart (1985) remarks that dropping this assumption would not matter for existence of an equilibrium as long as each firm's profit function is quasi–concave in its production plan. This quasi–concavity is not a too strong assumption since the functional form of subjective inverse demand functions expresses subjective beliefs of the firm, which can be assumed nice enough such that the decision problem has a unique solution.

3.2.3 Cournot–Walras equilibria

A Cournot–Walras equilibrium has been introduced by Gabszewicz & Vial (1972). In the framework of Chapter 2, it is an objective equilibrium of the dynamic system with strategic interaction as defined in 3.2.1 when all prices are strategic ($L_c = \emptyset$) and profit incomes adjust fast. This implies in particular that the outcome function does not depend on past outcomes, but only on actual production plans. Thus, one may write $A = \mathbb{R}^{\ell}_{++}$ and $F(y) = P(y) \in \mathcal{W}(y)$ with

$$\mathcal{W}(y) := \left\{ p \in \mathbb{R}^{\ell}_{++} \mid z_l(p, y) = 0 \ \forall\, l \in L \right\} \quad .$$

Definition 2: (y^*, p^*) is a *Cournot–Walras equilibrium* if $p^* = P(y^*)$ and if for all $j \in J$

$$y_j^* \in \operatorname*{argmax}_{y_j \in Y_j} P(y_j, y_{-j}^*) y_j \quad .$$

In the following, some of the problems with this definition, besides the problem of existence, will be reviewed shortly.

1. It can in general not be guaranteed that all production plans $y \in Y$ are *feasible*, i.e. that the set $\mathcal{W}(y)$ is nonempty, even if standard assumptions on consumers' preferences, initial endowments, and technologies are imposed. One assumption needed to prove existence of an equilibrium in the modified exchange economy is that the modified endowment vector is contained in the interior of the consumption set, i.e. $\omega_i(y) \gg 0$.[7] Gabszewicz & Vial (1972) achieve this by supposing $Y_j \subset \mathbb{R}^{\ell}_{+}$, but in general such an assumption constrains feasible production plans considerably. In particular, the set of feasible production plans of each single firm can depend on the choice of production plans of other firms. One might argue that it only makes sense for firms to take feasible production plans into account in their maximization problem, but this is very unsatisfactory, in particular if pure intermediate goods are introduced. For instance, if firm 1 uses for the production of a single output good a single intermediate good that is only produced by firm 2 but neither initially endowed nor consumed by consumers, the set of feasible production plans of firm 1 is trivially restricted by the production of firm 2, and vice versa. Thus, the assumption $Y \subset \{y \in \mathbb{R}^{n\ell} \mid \mathcal{W}(y) \neq \emptyset\}$ is in general very strong, but will still be imposed in the following.

2. Since $z(.,y)$ is homogeneous of degree zero, market clearing prices are defined only up to normalization. Thus $\mathcal{W}(y) \neq \emptyset$ implies also that $\mathcal{W}(y) \cap \Delta^{\ell} \neq \emptyset$, where $\Delta^{\ell} \subset \mathbb{R}^{\ell}_{+}$ denotes the standard $\ell - 1$–dimensional simplex. Denote by $\tilde{p} : Y \to \Delta^{\ell}$, $\tilde{p}(y) \in \mathcal{W}(y)$, a simplex normalized price selection, and by $\lambda : \Delta^{\ell} \to \mathbb{R}_{++}$ a real–valued function defining price normalization. That is, a general

[7] Using *resource relatedness* conditions, this assumption can be weakened, see Arrow & Hahn (1971), Chapter 5.

outcome function may be written

$$P(y) = \lambda(\tilde{p}(y)) \cdot \tilde{p}(y) \quad .$$

Two problems emerge. First, since there are non–pathological situations with multiple normalized equilibrium prices, the choice of the price function $\tilde{p}$ may require a selection among different equilibrium prices, and must have discontinuities if the equilibrium surface exhibits a fold. Dierker & Grodal (1986) show that such discontinuities can even lead to non–existence of equilibria in mixed strategies.

Second, the price normalization rule λ influences equilibrium allocations, as pointed out by Gabszewicz & Vial (1972). Böhm (1994) and Grodal (1996) prove theorems saying that arbitrary and therefore even inefficient production plans may be a Cournot–Walras equilibrium for specific price normalization rules. Moreover, some price normalization rules may lead to existence, others to non–existence of Cournot–Walras equilibria. Dierker & Grodal (1986) discuss the connection between the price selection and the normalization rule on existence of equilibria in pure and mixed strategies.

3. The dependence on price normalization may be attributed to the choice of profit maximization as an arbitrary objective of the firm which may not be in the interest of shareholders. Consider, for instance, the situation where each firm is owned by a single consumer only, i.e. $\theta_{ij} \in \{0, 1\}\ \forall i, j$. In this case it is natural to assume that each firm j maximizes instead of profits the indirect utility of its shareholder $i = i(j)$ with $\theta_{i(j)j} = 1$ defined by

$$u_i\Big(x_i(P(y_j, y^*_{-j}), P(y_j, y^*_{-j}) \cdot \omega_i(y_j, y^*_{-j}))\Big) \quad .$$

This objective function does not anymore depend on the choice of price normalization. But still, even for continuous selections $\tilde{p}$ objective equilibria need not exist, and discontinuous price selections may lead to non–existence of mixed strategy equilibria, too

(Dierker & Grodal (1986)). Apart from this, the choice of shareholder's indirect utility as a firm's objective function leads to some conceptual complications: First, shareholders behave schizophrenically since they act strategically as managers of their firm, but non–strategically as consumers on the market in the pure exchange economy. For instance, a shareholder might gain if he would not deliver all produced goods to the market, but would consume them directly. Second, this concept cannot be readily extended to the case of many shareholders in which a bargaining problem over production plans in the first stage would have to be solved, see also the discussion in Grodal (1996) and Dierker & Grodal (1996).

3.2.4 Cournot–Walras equilibria without income feedback

In a Cournot–Walras equilibrium each firm takes into account that its production decision influences profit incomes, and that this in turn has an effect on the demand function and therefore on the profit again. Such income effects are sometimes also called "Ford effects"[8]. It is, however, not always plausible to make such an assumption, and in later chapters other models will be investigated where these Ford effects are not taken into account by firms.

One may as well consider the case of an economy consisting of many markets ("islands") on which certain producers and consumers trade, but where consumers receive profit income[9] in other parts of the economy. A change of a single firm's production plan therefore influences in the first place the incomes of shareholders of this firm, but only indirectly (via prices and changes of profits of other firms) the profit incomes of

[8] After Henry Ford who is known for having realized a feedback effect of the wage payments on sales.

[9] In the interpretation of Gabszewicz/Vial also other factor incomes.

consumers in the home market. These indirect feedback effects are small if there is a large number of such islands and if profit shares are sufficiently spread over consumers, see Hart (1985). At least it can be supposed that these indirect effects take time, and adjust therefore, as in the above more simpler formulation, slowly from period to period so that firms reasonably take profit incomes in the home market as given and take only the adjustment of prices into account. In the following, all goods are again assumed to be strategic, i.e. their prices adjust immediately to the market clearing level.

Without loss of generality, the outcome set may be written $A := \mathbb{R}^{\ell}_{++} \times Y \times \mathbb{R}^{m}_{+}$ and a typical element is (p, y, π). Given $y \in Y$ and $\pi \in \mathbb{R}^{m}_{+}$, the inverse demand function is a selection of the set of market clearing prices

$$\mathcal{W}(y, \pi) = \left\{ p \in \mathbb{R}^{\ell}_{++} \mid \hat{z}_l(p, y, \pi) = 0 \ \forall \ l \in L \right\} .$$

As before, this set may be empty for some choices of production plans. At least, under the Gabszewicz/Vial assumption $Y_j \subset \mathbb{R}^{\ell}_{+}$, $\mathcal{W}(y, \pi)$ is nonempty for all $y \in Y$, $\pi \in \mathbb{R}^{m}_{+} \setminus \{0\}$ and if $\omega_i \gg 0 \ \forall i \in I$, since non–emptiness is equivalent to existence of a market clearing price in the pure exchange economy with endowments $\tilde{\omega}_i = \omega_i + \frac{\pi_i}{\bar{\pi}} \bar{y}$, where $\bar{\pi} := \sum_{i \in I} \pi_i$ and $\bar{y} := \sum_{j \in J} y_j$.

Since $\hat{z}$ and therefore $\mathcal{W}(y, \pi)$ depend on the vector of production plans $(y_j)_{j \in J}$ only via $\bar{y} = \sum_{j \in J} y_j$, a selection may be written $P(\bar{y}, \pi)$ and the outcome function is [10]

$$F : A \times S \ni \Big((\tilde{p}, \tilde{y}, \tilde{\pi}), y \Big) \mapsto (p, y, \pi) \in A \quad \text{with}$$

$$\pi = \Big(\textstyle\sum_{j \in J} \theta_{ij} \tilde{p} \tilde{y}_j \Big)_{i \in I} \quad \text{and} \quad p = P(\bar{y}, \pi)$$

The following definition is just a reformulation of an objective equilibrium:

[10] At least under the Gabszewicz/Vial assumption $Y_j \subset \mathbb{R}^{\ell}_{+}$ this function is well–defined.

Definition 3: (y^*, p^*, π^*) is a *Cournot–Walras equilibrium without income feedback* if

(i) $y_j^* \in \mathrm{argmax}_{y_j \in Y_j} P(y_j + \sum_{k \neq j} y_k^*, \pi^*) y_j \quad \forall\, j \in J$,

(ii) $p^* = P(\bar{y}^*, \pi^*)$,

(iii) $\pi_i^* = \sum_{j \in J} \theta_{ij} p^* y_j^* \quad \forall\, i \in I$.

An immediate consequence is that this equilibrium is independent of price (and profit) normalization: If (y^*, p^*, π^*) is an equilibrium, then also $(y^*, \lambda p^*, \lambda \pi^*)$ is an equilibrium for any $\lambda > 0$ since $P(\bar{y}, .)$ is linearly homogeneous.[11]

From the definition of the inverse demand function P and the budget constraints of consumers it follows that the sum of all firms' profits is independent of production plans and equals the aggregate profit income:

$$
\begin{aligned}
P(\bar{y}, \pi)\bar{y} &= P(\bar{y}, \pi)\Big(\sum_{i \in I} (x_i(P(\bar{y}, \pi), P(\bar{y}, \pi)\omega_i + \pi_i) - \omega_i)\Big) \\
&= \sum_{i \in I} \pi_i = \bar{\pi} \qquad \forall\, (y, \pi) \, . \qquad (3.2)
\end{aligned}
$$

Notice that $p = P(\bar{y}, \pi)$ is in general a selection of the Cournot–Walras correspondence $\mathcal{W}(y)$ only if the income distribution is correct, i.e. if $\pi_i = \sum_{j \in J} \theta_{ij} p y_j$ for all $i \in I$. This is clearly the case if there is only one consumer ($m = 1$). In another special case where all consumers have quasi–linear utility in one commodity, the distribution of profits plays no role and the inverse demand $P(\bar{y}, .)$ is a linear function of the aggregate profit income $\bar{\pi} = \sum_{i \in I} \pi_i$. In these two cases, $P(\bar{y}, \pi) \in \mathcal{W}(y)$ for all y and π, and a Cournot–Walras equilibrium without income feedback is a Cournot–Walras equilibrium under the specific price normalization defined by $p\bar{y} = \bar{\pi}$.

The case of a single monopolist ($n = 1$) turns out to be completely indeterminate: For any $y \in Y$, $(y, P(y, (\theta_i)_{i \in I}), (\theta_i)_{i \in I})$ is an equilibrium

[11] In fact, only $\mathcal{W}(y, .)$ is linearly homogeneous, and homogeneity of $P(\bar{y}, .)$ has to be assumed if $\mathcal{W}(y, .)$ is multi–valued.

since (3.2) implies that the monopolist's profit is constant, whenever the inverse demand function is defined. If there is more than one firm, the situation is different. (3.2) then implies that maximization of one firm's profit is equivalent to the minimization of aggregate profits of all other firms. A competition with this feature turns out to be efficient if all firms have identical technologies. The following theorem will show that whenever firms have identical technologies, a symmetric Cournot–Walras equilibrium without income feedback is a competitive equilibrium, and vice versa if profit functions are strictly quasi–concave. A *competitive equilibrium* will be defined here as a vector (y^*, p^*, π^*) which fulfills (ii) and (iii) in Definition 3, but satisfies instead of (i) the price–taking profit maximization condition

$$y_j^* \in \operatorname*{argmax}_{y_j \in Y_j} p^* y_j \quad \forall\, j \in J\ .$$

Theorem 1: Let $n > 1$, suppose $Y_j = \hat{Y}$ for all $j \in J$ where $\hat{Y} \subset \mathbb{R}^\ell$ is convex, let $(y^*, p^*, \pi^*) \in Y \times \mathbb{R}^\ell_{++} \times \mathbb{R}^m_+$ with $y^* = (\hat{y}^*)_{j \in J}$. Assume that $P(., \pi^*)$ is continuously differentiable in $n\hat{y}^*$. Then it follows:

(i) If (y^*, p^*, π^*) is a Cournot–Walras equilibrium without income feedback, then it is a competitive equilibrium.

(ii) If (y^*, p^*, π^*) is a competitive equilibrium and the function $\Pi : y \mapsto P(y + (n-1)\hat{y}^*, \pi^*)y$ is strictly quasi–concave, then it is a Cournot–Walras equilibrium without income feedback.

Proof: Define $A := d_y P(n\hat{y}^*, \pi^*) \in \mathbb{R}^{\ell \times \ell}$. Then differentiation of (3.2) at $n\hat{y}^*$ implies

$$0 = \frac{d}{dy}\Big(P(ny, \pi^*)ny\Big)\Big|_{y=\hat{y}^*} = n(A \cdot n\hat{y}^* + p^*) \quad ,$$

where $p^* = P(n\hat{y}^*, \pi^*)$. With $\Pi(y) := P(y + (n-1)\hat{y}^*, \pi^*)y$, this yields

$$d\Pi(\hat{y}^*) = A \cdot \hat{y}^* + p^* = \frac{n-1}{n} p^* \quad . \tag{3.3}$$

If (y^*, p^*, π^*) is a Cournot–Walras equilibrium without income feedback, then $\hat{y}^* \in \operatorname{argmax}_{y \in \hat{Y}} \Pi(y)$ and convexity of $\hat{Y}$ imply $d\Pi(\hat{y}^*)(y - \hat{y}^*) \leq 0$ for all $y \in \hat{Y}$. Using (3.3) gives $p^*\hat{y}^* \geq p^*y$ for all $y \in \hat{Y}$, and claim (i) follows.

If (y^*, p^*, π^*) is a competitive equilibrium, it follows again from (3.3) and $p^*\hat{y}^* \geq p^*y$ that $d\Pi(\hat{y}^*)(y - \hat{y}^*) \leq 0$ for all $y \in \hat{Y}$. But now strict quasi–concavity of $\Pi(.)$ and convexity of $\hat{Y}$ imply $\Pi(\hat{y}^*) \geq \Pi(y)$ for all $y \in \hat{Y}$, and therefore (ii) follows. □

The next counterexample will show that the results of Theorem 1 do not carry over to competition between firms with non–identical technologies.

Counterexample: Let $m = 1$, $n = 2$, $\ell = 2$ and assume that the single consumer has no endowment and that his preferences are represented by $u(x_1, x_2) = v(x_1) + v(x_2)$, where $v : \mathbb{R}_+ \to \mathbb{R}$ is strictly increasing, strictly concave, and fulfills $v(x) = x - x^2/2$ for all $0 \leq x \leq 3/4$. Technologies are

$$Y_j = \left\{(y_{j1}, y_{j2}) \mid 0 \leq y_{jj} \leq 3/4,\ y_{ji} = 0 \text{ if } i \neq j\right\} \quad , \ j = 1, 2 \ .$$

With abuse of notation, write y_j instead of y_{jj}.

The inverse demand function turns out to be

$$\begin{aligned} P_1((y_1, y_2), \pi) &= \frac{\pi v'(y_1)}{y_1 v'(y_1) + y_2 v'(y_2)} \ , \\ P_2((y_1, y_2), \pi) &= \frac{\pi v'(y_2)}{y_1 v'(y_1) + y_2 v'(y_2)} \ . \end{aligned}$$

Thus, maximization of $P_1((y_1, y_2), \pi)y_1$ over $y_1 \in [0, 3/4]$ and given $y_2 > 0$ is equivalent to the maximization of $y_1 v'(y_1) = y_1(1 - y_1)$ which has the solution $y_1 = 1/2$. By symmetry, the best response of firm 2 is $y_2 = 1/2$ and equilibrium prices are $p_1 = p_2 = \pi$. Thus, the unique Cournot–Walras equilibrium without income feedback differs from the unique Walrasian equilibrium with production plans $y_1 = y_2 = 3/4$ and prices $p_1 = p_2 = \pi/2$. ◇

According to this counterexample, the Bertrand–like result of Theorem 1 seems to hold only in very restricted cases where all firms have identical technologies. However, the result that firms behave as if they were price takers follows also if an identity like (3.2) holds for subsets of firms with equal technologies. For instance, suppose that there is a partition of goods in different sectors $L = \bigcup_{s\in S} L_s$, and that each firm produces only goods in one sector, but all firms in one sector have identical technologies. If all consumers have Cobb-Douglas preferences over subutilities of consumption in different sectors, they spend a fixed share of their income in each sector. If they have no endowments of goods produced by firms, an identity like (3.2) must be fulfilled in each sector and the result of Theorem 1 applies also to this case.

A Cournot–Walras equilibrium without income feedback is one particular modification of the definition of Cournot–Walras equilibria given by Gabszewicz and Vial. It coincides with a Cournot–Walras equilibrium under one specific price normalization rule in some special cases in which the income distribution has no influence on aggregate excess demand of consumers. There can be defined several other modifications of the Cournot–Walras approach in which firms take some prices as given, as it has been indicated in 3.2.1. Thus, there are several ways to define objective equilibria in this setup, depending on the choice of the outcome function F. In contrast, Negishi equilibria are defined independently of the particular choice of the outcome function F (i.e. whether some prices or the profit incomes adjust fast or slowly), but they depend heavily on the specification of subjective outcome functions F^j.

3.3 Price competition

3.3.1 Some preliminaries

Models with quantity competition as described above lack an explanation of price formation. Prices are either determined by an inverse demand function at their market clearing level, or they adjust according to a tâtonnement dynamics. In any case, they are set by an "invisible hand". In models with price competition at least some prices are controlled by agents in the economy and their determination is the result of decisions of these agents.

Goods can, as before, be distinguished between competitive and strategic goods $L = L^c \cup L^s$. It is assumed that $L^s = \bigcup_{j \in J} L_j$ where L_j is the set of goods whose prices are controlled by firm j. A good is therefore identified not only by its physical characteristic, but also by the agent who sets its price. Notice, however, that the assumption of strictly convex preferences excludes that two firms have price–setting power of goods which are identical in consumption, i.e. perfect substitutes for all consumers. Assume that $L_j \neq \emptyset$ for all $j \in J$, i.e. each firm has control over some prices. The case where some firms take all prices as given can be treated similarly but with a higher notational effort. Thus, the action set of firm j is $S^j = \mathbb{R}^{\ell_j}_{++}$, where $\ell_j = \sharp L_j$. Let $p^j = (p_l)_{l \in L_j} \in S^j$ denote a price vector chosen by firm j, let $p^s = (p^j)_{j \in J} \in R^{\ell^s}_{++}$ be a vector of all strategic goods' prices, and let $p^c \in \mathbb{R}^{\ell^c}_{++}$ be a vector of competitive goods' prices.

3.3.2 Equilibria with Walrasian demand

In the following, the equilibrium with price–setting firms introduced by Marschak & Selten (1974) and further developed by Hart (1985) will be described as an objective equilibrium of a specific dynamic system with strategic interaction. It characterizes a situation where (1) firms set prices, and (2) firms serve all demand addressed to them after they announced their prices.

Firms set prices of strategic goods simultaneously and take prices of competitive goods as given. As above, competitive prices adjust slowly in order to be determined independently of current decisions of firms. Profit incomes also adjust slowly, so that firms take them as given, as it is the case in the formulation of Hart (1985).

It is assumed that goods whose prices are controlled by one firm are neither produced nor used as inputs by other firms. Formally:

Assumption: $y_{jl} = 0$ for all $j \in J$ and $l \in L^s \setminus L_j$.

This assumption implies that the demand of a good addressed to the firm who controls its price is given by the demand of consumers alone, which is

$$\xi_l(p, \pi) := \sum_{i \in I} (x_{il}(p, p\omega_i + \pi_i) - \omega_{il}) \quad .$$

Define the outcome set as $A := \mathbb{R}^{\ell}_{++} \times Y \times \mathbb{R}^m_+$ with a typical element (p, y, π). Actions are $p^s = (p^j)_{j \in J} \in S = \mathbb{R}^{\ell^s}_{++}$. The outcome function may be written

$$\begin{aligned} F : A \times S &\rightarrow A \, , \\ \Big((\tilde{p}, \tilde{y}, \tilde{\pi}), p^s\Big) &\mapsto (p, y, \pi) \, , \end{aligned}$$

with the following definitions. Prices of competitive goods adjust according to a tâtonnement dynamics

$$p_l = G_l(\tilde{p}_l, \hat{z}_l(\tilde{p}, \tilde{y}, \tilde{\pi})) \quad , \; l \in L^c \, ,$$

where $\hat{z}_l$ and G_l are defined as in 3.2.1.

Consumers' profit incomes are given by the previous period's dividend payments

$$\pi_i = \sum_{j \in J} \theta_{ij} \tilde{p} \tilde{y}_j \quad .$$

Production plans y_j are solutions of

$$\max py \quad \text{s.t.} \quad y \in Y_j \;\text{ and }\; y_l = \xi_l(p, \pi) \quad \forall\, l \in L_j \quad . \tag{3.4}$$

Thus, each firm serves the Walrasian demand for those goods whose prices it controls, but decides freely about the production and factor usage of competitive goods. For instance, if the firm sets only the price of a single output good which is produced from several input goods, (3.4) expresses the usual cost minimization condition. There exists a unique solution of (3.4) if $\{y \in Y_j \mid y_l = \xi_l(p, \pi)\ \forall\, l \in L_j\} \neq \emptyset$, and if Y_j is closed, strictly convex and bounded above. Denote by $\Phi_j(p, \pi)$ a solution. Thus,

$$y_j = \Phi_j(p, \pi) \quad .$$

Subjective equilibria in this setting have, at least to my knowledge, never been defined in the literature, but it can be shown that Negishi equilibria are in fact also subjective equilibria of this system if the subjective inverse demand functions P_j (cf. Section 3.2.2) as mappings from $Y \cap \mathbb{R}^{\ell_j}$ to $\mathbb{R}^{\ell_j}_{++}$ are invertible with inverse D_j. In this case, condition (i) in Definition 1 is equivalent to

$$\begin{gathered} D_j(p^{j*}) = (\xi_l(p^*, \pi^*))_{l \in L_j} \quad \text{and} \\ (y_j^*, p^{j*}) \in \operatorname{argmax}\ (p^j, p^{-j*}) \cdot y_j \quad \text{s.t.} \\ y_{jl} = D_{jl}(p^j)\ \forall\, l \in L_j,\ y_j \in Y_j, \text{ and } p^j \gg 0 \quad . \end{gathered}$$

But this maximization condition is equivalent to (3.4) together with profit maximization with respect to prices and subjective demand functions D_j. Therefore, Negishi equilibria may also be regarded as subjective equilibria with price–setting agents.

An objective equilibrium of this dynamic system with strategic interaction turns out to be almost[12] identical to the imperfectly competitive equilibrium defined in Hart (1985), p. 117, and can be formulated as follows:

Definition 4: (y^*, p^*, π^*) is an *(objective) equilibrium with Walrasian demand* if

(i) $p^{j*} \in \operatorname{argmax}\left\{(p^j, p^{-j*}) \cdot \Phi_j((p^j, p^{-j*}), \pi^*) \mid p^j \in \mathbb{R}^{\ell_j}_{++}\right\} \ \forall\, j \in J,$

(ii) $\hat{z}_l(p^*, y^*, \pi^*) = 0 \quad \forall\, l \in L^c$,

(iii) $\pi_i^* = \sum_{j \in J} \theta_{ij} p^* y_j^* \quad \forall\, i \in I$.

Since all Φ_j are homogeneous of degree zero (this has to be assumed if there are multiple solutions of (3.4)), then it is easy to see that equilibria with Walrasian demand are independent of price normalization, i.e. if (y^*, p^*, π^*) is an equilibrium, then also $(y^*, \lambda p^*, \lambda \pi^*)$ is an equilibrium for any $\lambda > 0$.

Existence of an equilibrium has been proven by Hart (1985), but he has to make strong assumptions on the existence of a continuous best response function. Apart from the existence problem, this equilibrium concept can be criticized for different reasons. First, the assumption that each strategic good is in the production set of only one firm is very restrictive and excludes among other things all situations with intermediate goods. Second, and most importantly, the functions Φ_j defined above are in general not well–defined. This is due to the fact that the Walrasian excess demand $\xi(p, \pi)$ is typically not contained in the aggregate production set $\sum_{j \in J} Y_j$ for *all* prices and profits. For instance, if some price tends to zero and thereby the demand of this good tends to infinity, it may technologically not be possible for the firm to serve this demand. Moreover,

[12] Hart actually assumes that firms do not take into account the effects of the whole income of consumers on demand, i.e. also of non–profit income. This distinction does not matter if all consumers have only endowments of competitive goods.

even if it happens that one firm who changes its price is able to satisfy the Walrasian demand, other firms may not be able to do so. Therefore, some rationing must occur, and spillovers of demand between producers must be taken into account.

But even if all functions Φ_j would be defined on the whole price/profit–domain, the behavior of firms as supposed in (3.4) may be non–voluntary, if they can decide freely about production. For instance, in the single–input/single–output case with strictly increasing marginal costs, a firm is willing to satisfy demand only up to the "price equal to marginal cost" limit. Above this limit, profits are decreasing and can become negative for large output levels. Thus, at some price levels a firm may want to ration demand.

One might argue that such rationing never occurs in equilibrium since a firm who wants to ration demand could increase its price slightly while selling the same amount.[13] However, this statement holds true only in equilibrium, but provides no justification for the firms' behavior in other situations. But the behavior of firms outside the equilibrium is of particular importance in Definition 4, since the usage of Φ_j as objective demand functions is actually justified by the assumption that each firm satisfies demand *whatever* prices are set by all firms. Furthermore, it is a basic feature of price competition that firms announce their prices for a certain period over which they cannot change them. The above informal arbitrage argument does therefore not apply during such a period. Thus, if one firm changes its price, other firms might ration their demand as long as they are not forced to satisfy it, such that there are positive or negative spillover effects on the demand of the firm who changes the price. This implies that the Walrasian demand ξ_l does not capture the true demand of a single firm (see Benassy (1988) and Benassy (1991)).

[13]Hart (1985), footnote 15, provides a similar justification by giving an arbitrage argument on a perfect resale market. Madden & Silvestre (1991) show that trade is always voluntary in an equilibrium with Walrasian demand under the assumption that goods are gross substitutes.

A solution of this Edgeworth problem[14] in a general equilibrium model will be presented in the next section.

3.3.3 Equilibria with effective demand

In this subsection, the equilibrium concepts introduced by Benassy (1976) and Benassy (1988) will be discussed. They characterize situations where (1) firms set prices and (2) they may produce less than demanded. In order to describe situations with demand and/or supply rationing in a general equilibrium model, one has to make use of the so–called *fixprice theory* developed firstly in full rigor by Benassy (1975) and Drèze (1975). This theory describes how exchange can take place at non–Walrasian price systems given that trade is *voluntary* (i.e. no agent trades more of each good than he wants to) and satisfies the *short–side rule* (i.e. no rationing on the short side of each market). A fixprice equilibrium describes such allocations by specifying a *rationing scheme*, i.e. a list of functions mapping demands of agents on each market into their transactions. In equilibrium, the demand of each agent is a best response to the trading constraints each agent observes.

In the following, the equilibrium concept introduced by Benassy will be used.[15] Other approaches on imperfect competition in models with quantity rationing are Grandmont & Laroque (1976) and Hahn (1978). However, both provide more a justification of fixprice equilibria by subjective inverse demand curves than an explicit description of price formation. If at all these equilibria can be formulated in the framework of Chapter 2, they would fit in Section 3.2. Böhm, Maskin, Polemarchakis & Postlewaite (1983) consider a monopolist who sets prices and trading

[14] Named after Edgeworth (1897), who introduced price competition where firms do not have an infinite potential supply.

[15] For a more general treatment of fixprice theory, including the Drèze approach, see Böhm (1989), Chapter 2.

constraints and finds it profitable to ration consumers. Unlike this, in Benassy's model firms do not set trading constraints strategically.

In the following exposition, it is assumed $p_1 = 1$, i.e. commodity $l = 1$ is chosen as a numeraire commodity. Each of the other $\ell - 1$ commodities can be exchanged in only one market against good 1. Trading constraints, demands and supplies for a good $l > 1$ have thus to be interpreted as trading constraints, demands and supplies on the specific market for exchanging good l against good 1.

Let the index $h \in I \cup J$ denote an agent of the economy, i.e. a consumer or a producer. A *fixprice equilibrium* at a given price system $p = (p_l)_{l \in L}$ consists of a list of effective demands $(\tilde{z}_h)_{h \in I \cup J} \in \mathbb{R}^{(\ell-1)\times(m+n)}$, a list of transactions $(z_h^*)_{h \in I \cup J} \in \mathbb{R}^{(\ell-1)\times(m+n)}$, a list of trading constraints $(\underline{z}_h, \overline{z}_h)_{h \in I \cup J} \in (\mathbb{R}_- \times \mathbb{R}_+)^{(\ell-1)\times(m+n)}$, and profit incomes $(\pi_i)_{i \in I} \in \mathbb{R}_+^m$ which fulfill equations (3.5), (3.6), (3.7) and (3.8) defined as follows. Transactions depend on effective demands and constraints by

$$z_{hl}^* = \max(\underline{z}_{hl}, \min(\tilde{z}_{hl}, \overline{z}_{hl})) \quad \forall\, h \in I \cup J,\ l \in L \setminus \{1\}\ . \tag{3.5}$$

Trading constraints are determined from demands via

$$\underline{z}_{hl} = \underline{G}_{hl}((\tilde{z}_{h'l})_{h' \in I \cup J}) \quad \text{and} \quad \overline{z}_{hl} = \overline{G}_{hl}((\tilde{z}_{h'l})_{h' \in I \cup J}) \quad , \tag{3.6}$$

where the functions G_{hl} specify a rationing mechanism, i.e. for all demands $\tilde{z}$ the actual transactions resulting from (3.5) and (3.6) fulfill voluntary trade, the short–side rule, and market clearing, see e.g. Benassy (1976) for formal definitions. Effective demands are defined by[16]

$$\tilde{z}_{il} \in \operatorname{argmax}_l \left\{ u_i(\omega_i + z) \mid pz \le \pi_i,\ \underline{z}_{il'} \le z_{l'} \le \overline{z}_{il'}\ \forall\, l' \ne l \right\},\ i \in I,$$

$$\tilde{z}_{jl} \in \operatorname{argmax}_l \left\{ -pz \mid -z \in Y_j,\ \underline{z}_{jl'} \le z_{l'} \le \overline{z}_{jl'}\ \forall\, l' \ne l \right\},\ j \in J. \tag{3.7}$$

Thus, Benassy's effective demands on market l are the utility or profit maximizing trades, while taking into account the trading constraints on

[16] $\operatorname{argmax}_l\{*\}$ denotes the lth component of $\operatorname{argmax}\{*\}$.

all other markets $l' \neq l$. Consistency of profit incomes is expressed by

$$\pi_i = \sum_{j \in J} \theta_{ij}(-pz_j^*) \quad , \; i \in I. \tag{3.8}$$

Existence of a fixprice equilibrium for arbitrary strictly positive prices has been proven by Benassy (1975), if all production sets are strictly convex and utility functions are strictly quasi–concave. Under these assumptions also the transactions resulting from the effective demands (3.7) are feasible (i.e. in the individual budget or production sets) and maximize utility or profits of the decision problems when trading constraints on *all* markets are taken into account (this need not hold if the "strictly" is dropped). Let $\mathrm{FP}(p) = (\tilde{z}(p), z^*(p), \underline{z}(p), \overline{z}(p), \pi(p))$ denote a selection of the set of fixprice equilibria in dependence on $p \in \mathbb{R}^{\ell}_{++}$.

An outcome function may be defined by assigning fixprice equilibria to prices either determined by price–setting firms or in competitive markets. The difficulties of the Walrasian outcome function in 3.3.2 disappear here, since the outcome function is well defined on the whole domain of strictly positive prices, and since transactions are always voluntary for all agents.

Formally, the outcome set is $A = \mathbb{R}^{4(\ell-1)(m+n)+m+\ell}$ with a typical element

$$a = (\tilde{z}, z^*, \underline{z}, \overline{z}, \pi, p) \quad .$$

Commodities $l > 1$ are either competitive ($l \in L^c$) or strategic ($l \in L^s = \bigcup_{j \in J} L_j$). Competitive prices are again determined by tâtonnement, but now on the basis of aggregate effective demands $\tilde{z}_l := \sum_{h \in I \cup J} \tilde{z}_{hl}$:

$$p_l = G_l(\tilde{p}_l, \tilde{z}_l(\tilde{p})) \quad , \; l \in L^c \; ,$$

where $\tilde{p}$ is the previous period's price vector. This determines the vector of competitive prices $p^c = (p_l)_{l \in L^c}$. Given actions of price–setting firms p^s, the rest of the outcome function is determined by $\mathrm{FP}(p)$ with $p = (p^c, p^s)$. Payoff functions are $-pz_j^*(p)$.

Benassy (1976) defines subjective equilibria of this dynamic system with strategic interaction. In his model price–setting agents have expectations about how they can influence trading constraints with their price choice. Formally, each firm j expects price–dependent constraints on markets $l \in L_j$, i.e. perceived demand and supply functions, $(\underline{Z}_l^e(p_l), \overline{Z}_l^e(p_l))$, and expectations about prices and trading constraints on markets which it does not control $p^{-j,e} \in \mathbb{R}_{++}^{\ell-\ell_j}$, $(\underline{Z}_l^e, \overline{Z}_l^e)$, $l \in L \setminus L_j$, and solves the problem

$$\max_{p^j,z} -(p^j, p^{-j,e})z \qquad \text{s.t.}$$
$$-z \in Y_j,\ \underline{Z}_l^e \leq z_l \leq \overline{Z}_l^e \text{ if } l \notin L_j, \text{ and } \underline{Z}_l^e(p_l) \leq z_l \leq \overline{Z}_l^e(p_l) \text{ if } l \in L_j.$$

Such a specification of firms' expectations defines a particular IOF, and a price choice of the above decision problem corresponds to the best action response to this IOF. At a subjective equilibrium, the perceived demand/supply functions coincide with the trading constraints at this price vector. Benassy (1976) justifies the idea of perceived demand curves by the following statement:

> "*Such a subjective perception is clearly much more realistic than the assumption that the monopolist knows the 'true' demand curve facing him. So, each time the monopolist has to make a price decision, he re-estimates his perceived demand curve as a function of what he observes (notably his maximal possible sales) and then chooses the prices of the goods he controls in the light of this perceived demand curve and his technological possibilities.*"

Benassy (1976) describes first the dynamics under *naive expectations*. This means that price and constraint expectations are equal to their realization of the previous period, while perceived demand and supply functions are "*consistent with the data of the previous period,* " which means that perceived demands/supplies at the previous prices equal the previous trading constraints. He defines then a monopolistic equilibrium as a fixed point of the resulting dynamical system (2.1) with naive expectations. Thus, a monopolistic equilibrium is a subjective equilibrium with the above defined IOFs. As an extension of the model, he also describes a general learning dynamics, where firms use the information of the previous T periods. Here he imposes "*the consistency condition ... that, if the monopolist has observed the same price and perceived constraint for T periods, the perceived demand curve should go through the corresponding point.*" This condition is very similar to the consistency condition imposed in Section 2.4, and one should not expect to select among subjective equilibria with this condition alone. Benassy suspects nonetheless, with reference to other unspecified information $I_j(t)$, that "*if this information is quite complete, the perceived demand curve may, after a sufficient number of observations, get quite close to the 'true demand curve'.*" However, where this other information should come from is left open.

In a later paper, Benassy (1988) defines an objective equilibrium of the above specified dynamical system with strategic interaction (for the case $L^c = \emptyset$) and proves existence under the usual quasi–concavity assumption on profit functions. Formally, an objective equilibrium is defined by

Definition 5: p^* is an *(objective) equilibrium with effective demand* if

$$\text{(i)} \quad p^{j*} \in \operatorname{argmax}\Big\{ -(p^j, p^{-j*}) z_j^*(p^j, p^{-j*}) \mid p^j \in \mathbb{R}_{++}^{\ell_j} \Big\} \quad \forall\, j \in J\,,$$
$$\text{(ii)} \quad \tilde{z}_l(p^*) = 0 \quad \forall\, l \in L^c\,.$$

It should be noticed that subjective outcome functions as defined in Benassy (1976) are never objective, even if perceived demand and supply functions on the markets under control of the agents are correct, i.e. $\underline{Z}_l^e(p_l) = \underline{z}_{jl}(p_l, p_{-l}^*)$ and $\overline{Z}_l^e(p_l) = \overline{z}_{jl}(p_l, p_{-l}^*)$ for all p_l and $l \in L_j$, except under very special circumstances. This is due to the fact that in an objective equilibrium an agent also takes into account the reactions of constraints on markets whose prices are not under his control, while this is ruled out by the definition of subjective outcome functions in a subjective monopolistic equilibrium. Nevertheless, it seems natural that firms believe to influence constraints only on markets under their control. A compromise between the two concepts might be to define partially objective equilibria, where agents hold objective demand functions only on the markets under their control.

It is a weakness of the outcome function to use a fixed point argument by assigning fixprice equilibria to different price choices (this is of course also true for the Cournotian models in 3.2.1 where inverse demand curves have been defined by competitive equilibria of pure exchange economies). Behind such a fixed point stands itself the idea of a fast quantity tâtonnement (which corresponds in the excess supply regime of a macroeconomic model to the traditional Keynesian multiplier process), where agents face in each step certain trading constraints, express their corresponding effective demands, which determine the constraints in the next step, and so on. Benassy (1976) describes this infinitely fast process also explicitly in combination with the slower price adjustment process. What is implausible, however, is the assumption that price–setting agents are not able to change prices during the fast quantity tâtonnement. For instance, an agent who finds out that his expected demand for some good is lower than the actually observed demand would certainly want to increase the price of this good. Of course, it has to be assumed that prices have to be fixed for a certain period, but it is not clear why quantities adjust infinitely fast compared to prices (or, if adjustments should be endogenized, why menu costs are infinitely larger than quantity adjustment costs). Benassy (1976) remarks that the concepts "*. . . would also apply to the case where prices and quantities adjust together. Though, we should then replace the analysis of 'successive' equilibria by an explicit dynamic analysis.*"

It further seems to be a strong restriction that all production takes place after prices have been announced. On many markets it is more plausible to think of firms to produce (or to install capacities) first and to set prices afterwards. Thus, more sophisticated combinations of quantity and price competition, as they have already been studied in partial market frameworks, have to be investigated also in general equilibrium models. Exchange economies with simultaneous price and quantity competition have been studied in the literature on strategic market games (see e.g. Hurwicz (1979) and Dubey (1982)), which lead typically to competitive outcomes. However, these models describe only one–stage competition, i.e. prices and trade offers are set simultaneously, and an allocation rule determines transactions. For a recent contribution to two–stage price–quantity competition in a strategic market game which leads to non–Walrasian outcomes see Germano (1995).

3.4 Summary

In this chapter several general equilibrium models with quantity and price competition have been presented. Subjective equilibria have been defined in the literature, where often plausible assumptions have been imposed on expected outcome functions. Objective equilibria have been defined in order to avoid the large degree of arbitrariness of expected outcome functions. They turn out not always to be selections of the set of subjective equilibria, since objective outcome functions may display features which are not compatible with the assumptions imposed on subjective outcome functions in the definitions of subjective equilibria. This weakens additionally the justification of objective outcome expectations by a learning argument.

Chapter 4

Intertemporal macroeconomic equilibrium

> "*This self–referential aspect of economic systems gives rise to enormous theoretical problems of indeterminacy (i.e. multiple equilibria) when people's expectations are left as 'free variables' that are not restricted by economic theory. To fight that threat of indeterminacy, economists have embraced the hypothesis of rational expectations.*"
>
> A. Marcet and T.J. Sargent (1992)

4.1 Introduction

Macroeconomic models with imperfect competition have been investigated intensively over the past two decades. Most of these models abstract from aggregation issues and consider representative agent economies with three types of goods: labor, a consumption good, and fiat money. In contrast to some models which incorporate money by including end–of–period balances as an argument in the utility function, this chapter follows the temporary monetary equilibrium approach, as introduced by Hicks (1946) and Patinkin (1956) and developed in full rigor by

Grandmont (1983). In this framework, a store of value motive of money is derived from an explicit modeling of intertemporal decision making and of future price expectations.

According to the viewpoint of temporary equilibrium theory, trade takes place sequentially over time. When agents make decisions, they must forecast their future environment (mostly the future spot market prices) using the information from current and past states of the economy. A temporary equilibrium is a state of the economy in a given period, given these forecasts of agents. Grandmont (1982) distinguishes between a temporary Walrasian equilibrium, in which prices are flexible and all markets clear, and a temporary equilibrium with quantity rationing, in which prices are rigid and trade takes place at a fixprice equilibrium.

In an economy with imperfect competition, a temporary equilibrium can be defined as an objective equilibrium in a single period given the agents' expectations about future prices. The usage of the objective demand approach is compatible with most of the literature on macroeconomics with imperfect competition (for surveys see Silvestre (1993), Silvestre (1995) and Dixon & Rankin (1994)). Some of these models do not have a temporary equilibrium interpretation, but use nominal (e.g. Hart (1982)) or real (e.g. Blanchard & Kiyotaki (1987)) money balances as an argument of utility functions. Such utility functions can be interpreted as indirect utility functions of an intertemporal utility maximization problem by supposing expectation functions which are inelastic or unitelastic in the actual price.

Along a sequence of temporary equilibria, agents may make forecast errors. The special case of a sequence of temporary equilibria where agents forecast the future correctly will be termed generally *intertemporal equilibrium.* In perfectly competitive models such intertemporal equilibria are often characterized as solutions of some implicit equation linking equilibrium states of two or more subsequent periods. If this implicit equation fulfills some generic non–degeneracy condition, stationary equilibria are locally isolated. In contrast to these models, it turns out that under imperfect competition there is a continuum of stationary equilibria. The reason for this indeterminacy[1], which has been recognized first by Rankin (1992), is that in models with imperfect competition not only point expectations about future prices, but whole expectation functions are decisive since they determine the demand functions and therefore the payoff functions of imperfect competitors. In particular, since the elasticity of these demand functions determines the markup, the elasticity of price expectations turns out to be an important determinant of an intertemporal equilibrium.

It is important in the following to make a precise distinction between a *perfect foresight equilibrium*, in which agents forecast the future correctly along the equilibrium path, and a *rational expectations equilibrium*, in which agents would also make correct forecasts in (hypothetical) situations off the equilibrium path. Rankin remarks that determinate rational expectations equilibria could be defined in principle, and he computes such a determinate rational expectations equilibrium in another paper (Rankin (1997)). However, it will be shown here that Rankin's result rests on a special assumption and that one typically cannot get rid of the indeterminacy problem.

This chapter is organized as follows: In the next section, the concepts of perfect foresight and rational expectations equilibria will be defined in a general temporary equilibrium framework. In Section 4.3 an OLG version of Hart's (1982) economy with an imperfectly competitive goods market and a perfectly competitive labor market will be described.[2] It will be shown that the expectation effects described above lead to multiple

[1]This usage of the term *indeterminacy* differs from the usual definition in OLG models (cf. Chiappori & Guesnerie (1991)) stating that in each neighborhood of some (stationary) equilibrium there exists an infinite number of intertemporal equilibria. Here indeterminacy means that in each neighborhood of a stationary equilibrium there exists an infinite number of other stationary equilibria.

[2]A previous version of Section 4.3 was published in Kaas (1998).

perfect foresight equilibria with involuntary unemployment, i.e. unemployment even at a zero wage rate. This extends Schultz's (1992) result, who restricted his analysis to inelastic price expectations. In Section 4.4 it will be shown that even the stronger concept of rational expectations does generally not tie down unique rational expectations equilibria. These results will be derived from a simple characterization of rational expectations equilibria in the general framework, and they will be applied to intertemporal Cournot–Walras equilibria and to a market game between consumers in a one–good pure exchange overlapping generations economy.

4.2 Temporary and intertemporal equilibrium

Before defining specific equilibria, the concepts of temporary and intertemporal equilibria will be described in a general framework. There is a state space X whose elements contain all the relevant information about the allocation in a given period. There are agents (which shall be in all subsequent models overlapping generations of consumers living for two periods) who have to take intertemporal decisions and make forecasts of the state in the next period. A *forecast function* $\psi : X \to X$ expresses how the expectation of an agent about the future state x^e depends on the current state x, i.e. $x^e = \psi(x)$. A *temporary equilibrium correspondence* is a correspondence $\mathrm{T} : X^X \rightharpoonup X$, which expresses how the set of temporary equilibria $\mathrm{T}(\psi)$ depends on the forecast function ψ of agents (which are all assumed to have the same forecast function). In general, such a temporary equilibrium correspondence may depend also on past states or other variables, but for the current purpose this formulation is sufficient.

Assume that time starts at some initial date $t = 0$ and that agents may use all past and current information when they form their forecasts. Hence, a *full forecast function* of the agents in period t is a mapping $\Psi_t : X^{t+1} \to X$ with the meaning $x^e_{t+1} = \Psi_t(x_t, \ldots, x_0)$. The set of temporary equilibria in period t, given a full forecast function Ψ_t and the past states $(x_{t-1}, \ldots, x_0)$, is then $\mathrm{T}(\Psi_t(., x_{t-1}, \ldots, x_0))$.

Definition 1: $(x^*_t)_{t\geq 0}$ is a *sequence of temporary equilibria generated by* $(\Psi_t)_{t\geq 0}$ if for all $t \geq 0$

$$x^*_t \in \mathrm{T}\Big(\Psi_t(., x^*_{t-1}, \ldots, x^*_0)\Big) \quad .$$

The next definition imposes a perfect foresight condition on such sequences of temporary equilibria.

Definition 2: $(x^*_t)_{t\geq 0}$ is a *perfect foresight equilibrium* (to be abbreviated PFE), if there is a sequence of forecast functions $(\Psi_t)_{t\geq 0}$ generating $(x^*_t)_{t\geq 0}$ as a sequence of temporary equilibria such that for all $t \geq 0$

$$x^*_{t+1} = \Psi_t(x^*_t, \ldots, x^*_0) \quad .$$

A PFE $(x^*_t)_{t\geq 0}$ is *stationary* if $x^*_t = x^*_{t+1}$ for all $t \geq 0$.

Clearly, $x^*_t = x^*$, $t \geq 0$, is a stationary PFE if and only if there exists $\psi : X \to X$ such that $\psi(x^*) = x^*$ and $x^* \in \mathrm{T}(\psi)$.

As explained in the introduction of this chapter, expectations outside the equilibrium are an important determinant of a temporary equilibrium with imperfect competition. Definition 2 imposes no restriction on these expectations, and one might suspect that this is a reason for indeterminacy of PFE. A possibility to define "rationality" of forecast functions is to require that they forecast temporary equilibrium states *whatever* happened in the past. Formally:

Definition 3: The sequence of forecast functions $(\Psi_t)_{t\geq 0}$, $\Psi_t : X^{t+1} \to X$ for all $t \geq 0$, is called *rational*, if for all $\tau \geq 0$ and all $(x_\tau, \ldots, x_0) \in X^{\tau+1}$, the sequence of forecasts $(x_t)_{t>\tau}$ defined recursively by $x_t = \Psi_{t-1}(x_{t-1}, \ldots, x_0)$, $t > \tau$, describes temporary equilibria, i.e. it satisfies $x_t \in \mathrm{T}(\Psi_t(., x_{t-1}, \ldots, x_0))$ for all $t > \tau$. A PFE, which is generated by a rational sequence of forecast functions, is called *rational expectations equilibrium*, to be abbreviated REE.

As an illustration of these definitions, the following two examples show that in some cases a distinction between PFE and REE is superfluous, while in others REE are a true selection of the set of PFE.

Example 1:

Some perfectly competitive models (for instance, the standard one–good pure exchange OLG model) are described by a mapping $Z : X \times X \to \mathbb{R}^n$, where $X \subset \mathbb{R}^n$ (e.g. Chiappori & Guesnerie (1991)). A temporary equilibrium x in a given period with expectations x^e for the next period is defined by the equation $Z(x, x^e) = 0$. An *intertemporal equilibrium* (IE) in such a model is defined as a sequence $(x_t^*)_{t\geq 0}$ fulfilling $Z(x_t^*, x_{t+1}^*) = 0$ for all $t \geq 0$ (see e.g. Grandmont (1988)). Thus, an intertemporal equilibrium is defined without using the concept of forecast functions. In the above framework, the temporary equilibrium correspondence is defined by $\mathrm{T}(\psi) = \{x \in X \mid Z(x, \psi(x)) = 0\}$. It is clear that there are many forecast functions generating a given intertemporal equilibrium as a PFE in the sense of Definition 2. Moreover, each intertemporal equilibrium is a REE in the sense of Definition 3. Given an IE $(x_t^*)_{t\geq 0}$, the sequence of forecast functions $\Psi_t(x_t, \ldots, x_0) := x_{t+1}^*$ is rational and generates $(x_t^*)_{t\geq 0}$ as a REE. Thus, IE $\subset$ REE $\subset$ PFE. On the other hand, each PFE $(x_t^*)_{t\geq 0}$ clearly fulfills $Z(x_t^*, x_{t+1}^*) = 0$ implying REE $\subset$ PFE $\subset$ IE. Therefore, the sets of PFE and of REE both coincide with the set of intertemporal equilibria, and out–of–equilibrium expectations do not matter. ◇

Example 2:

Consider a model with two–period living overlapping generations of consumers and two firms, each producing at zero cost up to one unit of a consumption good. Consumers consume in both periods of their life, have no endowment, and receive all profit income when they are young. Assume as Rankin (1997) that there are separated markets for young and for old consumers. Firm 1 is a monopolist on the market for young consumers, and firm 2 on the market for old consumers. Each old consumer holds an amount $M > 0$ of fiat money, and assume that firm 2 sets in each period the market clearing price $\bar{q} = M$ (actually firm 2 is indifferent between all prices $q \geq \bar{q}$). Assuming homothetic preferences, the young consumer's consumption demand can be written (see Section 4.3 below) $c(\frac{q^e}{p})\frac{I}{p}$ with a propensity to consume $0 < c(.) < 1$, where p is the price of the consumption good, I is his income, and q^e is the expected price for the next period. Income consistency can be expressed $I = c(\frac{q^e}{p})I + M$, implying that the income consistent demand of the young consumer is

$$D(p, q^e) := \frac{Mc(\frac{q^e}{p})}{(1 - c(\frac{q^e}{p}))p} \quad .$$

Assume that c is strictly increasing, i.e. that consumption in both lifetime periods are gross substitutes. This implies that maximization of firm 1's profit $D(p, q^e)p$ is equivalent to maximization of q^e/p. Let $\psi(p) = q^e$ be a price expectation function of the young consumer. Then firm 1's decision problem is

$$\max_{p>0} \frac{\psi(p)}{p} \quad \text{s.t.} \quad D(p, \psi(p)) \leq 1 \quad . \tag{4.1}$$

The state space X is the set of prices $(p, q) \in \mathbb{R}^2_{++}$. With abuse of notation, identify ψ with the second component of a full expectation function $\psi : X \to X$. Then the temporary equilibrium correspondence is

$$\mathrm{T}(\psi) := \{(p, q) \mid p \text{ solves (4.1) and } q = \bar{q}\ \} \quad .$$

Observe first that PFE are completely indeterminate. For each p_t satisfying $D(p_t, \bar{q}) \leq 1$, there exists a $\psi_t : \mathbb{R}_{++} \to \mathbb{R}_{++}$ such that $\psi_t(p_t) = \bar{q}$ and $p_t \in \operatorname{argmax}_{p>0} \psi_t(p)/p$. Thus, $(p_t, \bar{q}) \in \mathrm{T}(\psi_t)$, and each sequence $(p_t, \bar{q})_{t\geq 0}$ satisfying $D(p_t, \bar{q}) \leq 1$ for all t is a PFE.

In contrast, there exists only one REE. Since the only rational sequence of forecast functions $(\Psi_t)_{t\geq 0}$ must forecast the price $\bar{q}$ irrespectively of the past, the only solution of (4.1) with $\psi(.) \equiv \bar{q}$ is the unique $\bar{p}$ determined by $D(\bar{p}, \bar{q}) = 1$. Thus, the only REE is the constant sequence $(\bar{p}, \bar{q})_{t\geq 0}$.$\diamond$

4.3 Indeterminacy of perfect foresight equilibria and involuntary unemployment

4.3.1 Hart's model and involuntary unemployment

The article of Hart (1982) is the first which considers a macroeconomic model with imperfect competition and objective demand. Hart considers a static three–good model, where besides labor and a produced consumption good there is a nonproduced good whose role will be taken over by fiat money in the OLG version of the model. Prices are normalized with respect to the nonproduced good which is assumed to be consumed by and endowed to households only.

Hart assumes that there is a large number of identical households and firms which are assigned to different output and labor markets. This assignment of agents to markets and the distribution of profit shares is supposed to be such that each firm has a negligible influence on the incomes of consumers in its output market and on the wage rate in its labor market. On each labor market there is a certain number of unions, each having the same number of households as members. The number of firms in each output market and the number of unions in each labor market are parameters which measure the degree of competitiveness in these markets.

Firms are Cournot oligopolists on their output market, and unions are Cournot oligopolists on their labor market. The equilibrium considered by Hart is a Nash equilibrium of a two–stage game, where in the second stage firms set their output levels, given the wage rate and the income of consumers. In the first stage, unions set their labor supply knowing the inverse labor demand curve resulting from the second stage game and given the income of consumers. The income is determined by income consistency.

Since there is imperfect competition on the supply side of each market, prices exceed marginal costs, and wage rates are greater than or equal to their competitive level. Thus, the overall equilibrium is demand constrained. There may be unemployment or full employment, depending on the degree of competitiveness in these markets. It is not surprising that unemployment may occur if there are only few unions on each labor market, since unions are interested in the labor income of their members and tend to raise wage rates above the market clearing level.

However, it is at the first sight surprising that there can be unemployment even if the labor market is perfectly competitive, but the output market is not. Assuming no disutility of labor, it may happen that there is unemployment even at a zero wage rate, and one speaks in this case of *involuntary unemployment.*[3]

Such a situation can be illustrated in a partial market framework. Figure 4.1 shows the demand curve and the marginal revenue and marginal cost curves of a monopolist in a single output market. Since the marginal

[3] Actually, this type of unemployment is voluntary, since households supply at a zero wage rate any amount between zero and their labor endowment, but at any arbitrary small positive wage rate unemployment would be involuntary.

revenue curve intersects the abscissa in y_0, the monopolist never wants to supply more than y_0 for all wage rates w. If there is no disutility of labor, and if the labor endowment is positive for all wage rates such that y_0 is below the full employment output level $\bar{y}$, labor demand is lower than labor supply for all wage rates, even for $w = 0$. With a perfectly competitive goods market such a situation clearly cannot occur, provided that goods demand becomes greater than $\bar{y}$ for $p \to 0$.

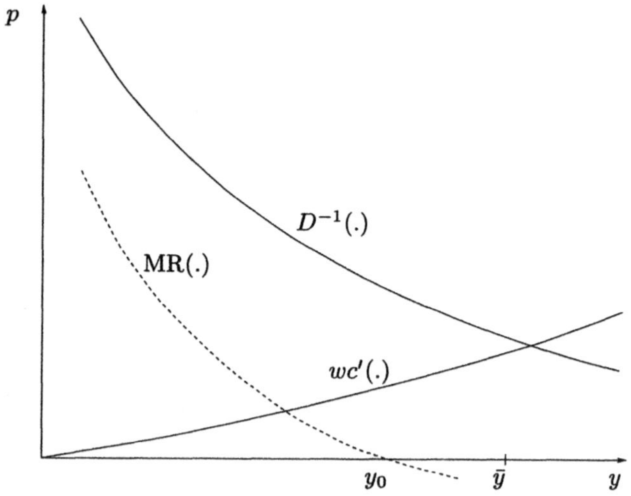

Figure 4.1: The monopolistic output does not exceed y_0 for all wage rates.

The critical issue is whether in Hart's general equilibrium model unemployment is necessarily due to market power of unions, or if also involuntary unemployment can occur. Hart himself excludes this type of unemployment simply by assuming that marginal revenue is always positive. Silvestre (1990) establishes, on the other hand, conditions under which marginal revenue becomes negative and he shows that these conditions may be fulfilled if households have CES utility over consumption plans with elasticity of substitution less than one.

Other authors show that involuntary unemployment may occur in modifications of Hart's economy in which, because of a different assumption on income determination, producers take the full income feedback on demand into account. Dehez (1985) shows the possibility of involuntary unemployment in a temporary monopolistic equilibrium version of such a model under very general conditions on price expectations and preferences. D'Aspremont, Dos Santos Ferreira and Gérard–Varet (1989 and 1991) extend this result to a static oligopolistic model and to an overlapping generations model.

Schultz (1992) considers an overlapping generations model in Hart's framework where firms do not take into account the income feedback on demand. He shows that involuntary unemployment cannot occur in any "rational expectations equilibrium", however under a very restrictive definition of rational expectations. That is, in the framework of 4.2, he only considers forecast functions which do not depend on the actual price, and he also remarks that his result depends crucially on the inelasticity of price expectations (p. 69/70).

In the following it will be shown, under very general conditions on the economy, that there are other intertemporal equilibria with involuntary unemployment. There exists a continuum of stationary PFE with unemployment at arbitrary low real wages, and within a certain interval nearly arbitrary perfect foresight trajectories are possible.

In the next subsection, temporary equilibria will be defined in an OLG version of Hart's model, and in 4.3.3 the main theorems on the indeterminacy of PFE will be proven. The results will be illustrated in 4.3.4 by an example, where a simple class of constant elasticity price expectations (containing log-linear adaptive and finite memory expectations) leads to a continuum of unemployment steady states. Additionally, the corresponding temporary equilibrium dynamics shows that unemployment equilibria may be stable, while full employment equilibria are not.

4.3.2 The economy

Consider an economy which evolves in discrete time periods $t = 0, 1, 2, \ldots$. In each period there are three commodities: labor, a consumption good and money. There are $n \geq 2$ identical firms producing the consumption good by means of labor input. The price of the consumption good is p_t, the wage rate is w_t and the price of money is normalized to one. Wage rates are taken as given by all agents in the economy, but their determination is left undescribed.[4] In the case of a competitive labor market, they are at the labor market clearing level, which is zero if there is involuntary unemployment. On the goods market there is Cournot competition.

The household sector is described by an overlapping generations structure, namely one representative young and one old agent in each period. The young household supplies inelastically a fixed amount of labor $\bar{L} > 0$ and receives wage as well as profit income from all firms. Households have preferences over consumption plans (c_1, c_2) in both periods of their life described by an intertemporal utility function $u : \mathbb{R}^2_+ \to \mathbb{R}$.

Assumption 1: u is C^2, homothetic, strictly quasi-concave and strictly monotonic on $\mathbb{R}^2_{++}$. $u'_i(c_1, c_2) \to \infty$ if $c_i \to 0$, $c_j > 0$, $j \neq i$, for $i = 1, 2$.

This implies that the utility maximization problem of a young household at income $I > 0$, price $p > 0$ and expected price $p^e > 0$ leads to a first-period consumption demand

$$c_1^* = c\left(\frac{p^e}{p}\right)\frac{I}{p}$$

[4]In Hart's model wages are set by unions in a first stage of each period. Such an assumption requires, however, the existence of a well–behaved eventual demand for labor curve describing the contingent solution of the second stage game, and thereby stronger assumptions on the economy.

with a continuously differentiable propensity to consume $c : \mathbb{R}_{++} \rightarrow (0, 1)$.

For a full description of the objective demand curve, price expectations have to be specified. Assume as in 4.2 that price expectations are described by a sequence of continuous forecast functions $\Psi = (\Psi_t)_{t \geq 0}$, where each $\Psi_t : \mathbb{R}^{t+1}_{++} \rightarrow \mathbb{R}_{++}$ maps current and past prices $(p_t, \ldots, p_0)$ into the for $t+1$ expected price p^e_{t+1}. As long as there is no reason for confusion, write $\psi_t(.) : \mathbb{R}_{++} \rightarrow \mathbb{R}_{++}$ instead of $\Psi_t(., p_{t-1}, \ldots, p_0)$.

The old household in period 0 holds an amount $M > 0$ of money. There is no government activity such that money market clearing implies that M is the stock of money held by all subsequent old generations. In every period the aggregate demand for the consumption good at young household's income I and forecast function ψ is given by the continuous function

$$D_\psi(p, I) := c\Big(\frac{\psi(p)}{p}\Big)\frac{I}{p} + \frac{M}{p} \quad .$$

It is easy to see that

$$\lim_{p \to 0} D_\psi(p, I) = \infty \qquad \text{and} \qquad \lim_{p \to \infty} D_\psi(p, I) = 0 \quad ,$$

so that there exists a (not necessarily unique and continuous) inverse mapping $P_\psi(., I)$, i.e.

$$P_\psi(Y, I) \in \{p > 0 \mid Y = D_\psi(p, I)\} \quad \forall\, Y > 0,\ I > 0 \;.$$

The n firms produce the consumption good from labor with identical labor requirement function $l : \mathbb{R}_+ \rightarrow \mathbb{R}_+$ which fulfills

Assumption 2: l is C^2 and satisfies $l' > 0$, $l'' \geq 0$ and $l(0) = 0$.

Firms act as Cournot oligopolists against the objective inverse demand $P_\psi(., I)$, and are assumed to take the wage rate and the income as given, i.e. they ignore that their strategic choice affects the income of the young consumer. Such an assumption can be justified by a sector structure of the economy where each household who receives wage and profit income in a sector consumes in some other sector and with consumers being uniformly distributed.

Before defining a temporary symmetric Cournot equilibrium, denote by

$$\hat{y}(y) := l^{-1}(\bar{L} - (n-1)l(y))$$

the maximum output of a single firm who takes the demand constraint on the labor market $\bar{L}$ and the output of all the other firms y as given.

Definition 4: Let $w \geq 0$ and $\psi \in C(\mathbb{R}_{++})$. A *temporary symmetric Cournot equilibrium relative to* (w, ψ) is a tuple (y, p, I) such that

$$\text{(i)} \quad y \in \operatorname{argmax}\{y' P_\psi((n-1)y + y', I) - wl(y') \mid 0 \leq y' \leq \hat{y}(y)\} \quad ,$$

$$\text{(ii)} \quad p = P_\psi(ny, I) \quad ,$$

$$\text{(iii)} \quad I = npy \quad .$$

The set of all temporary symmetric Cournot equilibria relative to (w, ψ) is denoted $\mathrm{TSCE}(w, \psi)$.

Notice that the concentration on symmetric equilibria is not too restrictive, since all unemployment equilibria must be symmetric if P_ψ is differentiable with a negative slope at any aggregate equilibrium output $Y = \sum_{j=1}^n y_j$. The first order conditions $y_j P'_\psi(Y, I) + P_\psi(Y, I) = wl'(y_j)$, $j = 1, \ldots, n$, can only be fulfilled if all y_j are identical, since the left hand side is strictly decreasing in y_j, while the right hand side is non–decreasing.

If ψ is differentiable, an easy computation yields

$$\frac{D'_\psi(p, I)p}{D_\psi(p, I)} = -1 + \frac{c(\frac{\psi(p)}{p})I\left(1 - c(\frac{\psi(p)}{p})\right)\left(\sigma(\frac{\psi(p)}{p}) - 1\right)\left(\eta_\psi(p) - 1\right)}{c(\frac{\psi(p)}{p})I + M} , \tag{4.2}$$

where

$$\sigma(\theta) := \frac{d}{d\theta}\left(\frac{\theta c(\theta)}{1-c(\theta)}\right)\frac{1-c(\theta)}{c(\theta)} = 1 + \frac{c'(\theta)\theta}{c(\theta)(1-c(\theta))} \geq 0$$

is the intertemporal elasticity of substitution and $\eta_\psi(p) := \psi'(p)p/\psi(p)$ is the elasticity of the forecast function. Since in a temporary equilibrium the income is determined by $I = M/(1 - c(\psi(p)/p))$ one gets

$$\frac{D'_\psi(p,I)p}{D_\psi(p,I)} = -1 + c\left(\frac{\psi(p)}{p}\right)\left(1 - c\left(\frac{\psi(p)}{p}\right)\right)\left(\sigma\left(\frac{\psi(p)}{p}\right) - 1\right)\left(\eta_\psi(p) - 1\right) . \tag{4.3}$$

Therefore, D_ψ and hence P_ψ are differentiable with negative slope in a temporary equilibrium, if, for instance, $(\sigma(\theta) - 1)(\eta_\psi(p) - 1) < 4$ uniformly in θ and p.

For $(y, p, I) \in \mathrm{TSCE}(w, \psi)$ and a continuously differentiable inverse demand function, the first order condition of (i) in Definition 4 is

$$p\left(1 + \frac{D_\psi(p,I)}{npD'_\psi(p,I)}\right) \geq wl'(y) \quad \text{with equality if} \quad nl(y) < \bar{L}.$$

Using (4.3), this is equivalent to

$$1 - c\left(\frac{\psi(p)}{p}\right)\left(1 - c\left(\frac{\psi(p)}{p}\right)\right)\left(\sigma\left(\frac{\psi(p)}{p}\right) - 1\right)\left(\eta_\psi(p) - 1\right) \geq \frac{p}{n(p - wl'(y))} \quad \text{with equality if} \quad nl(y) < \bar{L}. \tag{4.4}$$

From this condition an extension of the result of Schultz (1992, Theorem 1) on the impossibility of involuntary unemployment in a temporary equilibrium can be derived. If $(\sigma(\theta) - 1)(\eta(\theta) - 1) < 2$ for all expected inflation factors $\theta > 0$ (which holds under the Schultz assumption $\eta(.) \equiv 0$), there must be full employment for sufficiently low wages, since the left hand side of (4.4) is greater than $1/2$, while the right hand side converges to $1/n \leq 1/2$ for $w \to 0$. Therefore, the inequality must be strict for low wages which requires full employment. On the other hand, it seems clear that for arbitrary preferences and forecast functions (4.4) may be fulfilled with equality, and hence unemployment at arbitrary low wages can not be excluded. That this phenomenon can be consistent with perfect foresight will be shown in the next subsection. In accordance with Definition 2, perfect foresight equilibria and stationary equilibria are defined as follows:

Definition 5: Let $(w_t)_{t\geq 0} \geq 0$ be a sequence of non-negative wages. A sequence $((y_t, p_t, I_t))_{t\geq 0}$ is a *perfect foresight symmetric Cournot equilibrium relative to* $(w_t)_{t\geq 0}$ if there exists a sequence of forecast functions $\Psi = (\Psi_t)_{t\geq 0}$, $\Psi_t \in C(\mathbb{R}^{t+1}_{++}, \mathbb{R}_{++})$ $\forall t \geq 0$, such that for all $t \geq 0$

$$\begin{array}{ll} \text{(i)} & (y_t, p_t, I_t) \in \text{TSCE}(w_t, \Psi_t(., p_{t-1}, \ldots, p_0)) \quad , \\ \text{(ii)} & \Psi_t(p_t, \ldots, p_0) = p_{t+1} \quad . \end{array}$$

Definition 6: A *stationary symmetric Cournot equilibrium relative to* $w \geq 0$ is a tuple (y, p, I) such that there exists $\psi \in C(\mathbb{R}_{++})$ with

$$\begin{array}{ll} \text{(i)} & (y, p, I) \in \text{TSCE}(w, \psi) \quad , \\ \text{(ii)} & \psi(p) = p \quad . \end{array}$$

4.3.3 Indeterminacy of perfect foresight equilibria

It will be shown first that fixing the expected inflation factor θ is not sufficient to obtain unique temporary symmetric Cournot equilibria. Instead, a continuum of equilibria with unemployment can occur at arbitrary low wages. The only required assumption is an intertemporal elasticity of substitution different from 1, i.e. expectations have to matter locally.

Theorem 1: Let $\theta > 0$ and $\sigma(\theta) \neq 1$. Then there exist $\bar{w} > 0$ and $\tilde{y} < \bar{y} := l^{-1}(\bar{L}/n)$, such that for all $y \in [\tilde{y}, \bar{y}]$, $p = \dfrac{M}{ny(1-c(\theta))}$, and all $0 \leq w \leq \bar{w}$, there exists $\psi \in C(\mathbb{R}_{++})$ with $(y, p, npy) \in \text{TSCE}(w, \psi)$ and $\psi(p) = \theta p$. Moreover, if u is C^3, then $\bar{w}$ and $\tilde{y}$ depend continuously on θ.

Proof: $\sigma(\theta) \neq 1$ implies $c'(\theta) \neq 0$, and there exist $\underline{c} < c(\theta) < \bar{c}$, $\underline{\theta} < \theta < \bar{\theta}$ such that c maps $[\underline{\theta}, \bar{\theta}]$ diffeomorphically onto $[\underline{c}, \bar{c}]$. If u is C^3, c is C^2

and $\underline{c}$ and $\overline{c}$ may be chosen to depend continuously on θ (see Abraham, Marsden, and Ratiu [1, p. 105/106]). Define

$$\begin{aligned}
\overline{w} &:= M(1-1/n)\Big(n\overline{y}l'(\overline{y})(1-c(\theta))\Big)^{-1} \quad , \\
\tilde{y} &:= \min\left\{ y \geq 0 \;\middle|\; \begin{array}{ll} \frac{n-1}{n} + \frac{1}{n}(\frac{\hat{y}(y)}{y})^{(n-1)/n} & \leq \; 1-c(\theta)+\overline{c} \\ \frac{n-1}{n}\frac{y}{\hat{y}(y)} + \frac{1}{n} & \geq \; 1-c(\theta)+\underline{c} \end{array} \right\} .
\end{aligned}$$

Notice that $\tilde{y} < \overline{y}$ since both inequalities hold strictly for $y = \overline{y}$, and that $\tilde{y}$ is continuous in θ if $\underline{c}$ and $\overline{c}$ are continuous in θ, since $y \mapsto \hat{y}(y)/y$ is strictly monotonically decreasing.

Let $0 \leq w \leq \overline{w}$ and $\tilde{y} \leq y \leq \overline{y}$. Set

$$p := \frac{M}{ny(1-c(\theta))} \qquad \text{and} \qquad \eta := 1 - \frac{wl'(y)}{p} \geq \frac{1}{n} \quad .$$

Define $f : [0, \hat{y}(y)] \to \mathbb{R}_{++}$ by

$$f(y') := \begin{cases} \dfrac{npy}{y' + (n-1)y} & , \quad 0 \leq y' \leq y \quad , \\ \dfrac{py}{y'^{\eta} y^{1-\eta}} & , \quad y \leq y' \leq \hat{y}(y) \quad . \end{cases}$$

Clearly, f is continuous, strictly monotonically decreasing and $f(y) = p$. Define $\underline{d} := 1 + \underline{c}(1-c(\theta))^{-1}$ and $\overline{d} := 1 + \overline{c}(1-c(\theta))^{-1}$. It will be shown that

$$\underline{d}\frac{M}{y' + (n-1)y} \leq f(y') \leq \overline{d}\frac{M}{y' + (n-1)y} \qquad \forall\, y' \in [0, \hat{y}(y)] \quad .$$

Both inequalities clearly hold for $y' \leq y$ since $\underline{d}M < npy = M(1 - c(\theta))^{-1} < \overline{d}M$. For $y \leq y' \leq \hat{y}(y)$ the inequalities follow from the estimations

$$\begin{aligned}
\frac{(n-1)y + y'}{ny^{1-\eta}y'^{\eta}} &\geq \frac{n-1}{n}\frac{y}{\hat{y}(y)} + \frac{1}{n} \\
&\geq 1 - c(\theta) + \underline{c} = \underline{d}(1-c(\theta)) = \underline{d}\,\frac{M}{npy} \quad \text{and} \\
\frac{(n-1)y + y'}{ny^{1-\eta}y'^{\eta}} &\leq \frac{n-1}{n} + \frac{1}{n}\left(\frac{\hat{y}(y)}{y}\right)^{1-1/n} \\
&\leq 1 - c(\theta) + \overline{c} = \overline{d}(1-c(\theta)) = \overline{d}\,\frac{M}{npy} \quad .
\end{aligned}$$

Next it will be shown that y maximizes $\pi(y') := f(y')y' - wl(y')$ on $[0, \hat{y}(y)]$. Denote by π'_- (π'_+ resp.) the left (right resp.) derivative of π. The claim follows from

$$\begin{aligned}
\pi'_-(y') &= f(y')(1 - \frac{y'}{(n-1)y + y'}) - wl'(y) \\
&\geq p(1 - 1/n) - \bar{w}l'(\bar{y}) \\
&= \frac{(n-1)M}{n^2(1-c(\theta))}(\frac{1}{y} - \frac{1}{\bar{y}}) \geq 0 \quad , y' \leq y, \\
\pi'_+(y') &= f(y')(1-\eta) - wl'(y') \\
&\leq f(y)(1-\eta) - wl'(y) = 0 \quad , y' \geq y \, .
\end{aligned}$$

Choose any strictly monotonically decreasing $P \in C(\mathbb{R}_{++})$ such that

$$\begin{aligned}
\underline{d}\tfrac{M}{Y} \leq P(Y) \leq \bar{d}\tfrac{M}{Y} \qquad & \forall\, Y > 0 \quad \text{and} \\
P(Y) = f(Y - (n-1)y) \qquad & \forall\, Y \in [(n-1)y, (n-1)y + \hat{y}(y)] \quad .
\end{aligned}$$

P has a continuous inverse D satisfying

$$\underline{d}\tfrac{M}{p} \leq D(p) \leq \bar{d}\tfrac{M}{p} \qquad \forall\, p > 0 \quad .$$

With the definition

$$\psi(p') := p'c^{-1}\Big((1 - c(\theta))(p'D(p')/M - 1)\Big) \quad , p' > 0,$$

it follows $D_\psi(., npy) = D(.)$ and $\psi(p) = \theta p$. Furthermore, y is a symmetric Cournot equilibrium against the inverse demand $P_\psi(., npy) = P(.)$ and $P(ny) = p$. Hence, $(y, p, npy) \in \text{TSCE}(w, \psi)$. □

The central part of the proof of Theorem 1 is the construction of a forecast function yielding the existence of the desired temporary equilibrium. This forecast functions may be quite arbitrary, since there are, apart from continuity, no restrictions imposed on forecast functions. Existence of a Cournot equilibrium needs a well–behaved demand curve which requires in this model both restrictive assumptions on preferences and on price expectations. Since this theorem holds under general conditions on preferences, it is a technical necessity to allow for larger classes of forecast functions. It might be that another version of Theorem 1 can be proven using more restrictive classes of forecast functions while imposing stronger conditions on u. In the next subsection it will be shown that the same indeterminacy phenomenon occurs for the CES utility function and a class of constant elasticity forecast functions.

The next two corollaries apply Theorem 1 to get indeterminacy of perfect foresight and stationary equilibria.

Corollary 1: Let $K \subset \{\theta > 0 \mid \sigma(\theta) \neq 1\}$ be compact and assume u to be C^3. Then there exist $\bar{w} > 0$ and $\tilde{y} < \bar{y}$, such that for all $(w_t)_{t \geq 0}$ with $0 \leq w_t \leq \bar{w} \; \forall \, t \geq 0$ all sequences $((y_t, p_t, np_t y_t))_{t \geq 0}$ satisfying

$$y_t = \frac{M}{np_t(1 - c(\frac{p_{t+1}}{p_t}))} \in [\tilde{y}, \bar{y}] \quad \text{and} \quad \frac{p_{t+1}}{p_t} \in K \quad \forall \, t \geq 0$$

are perfect foresight symmetric Cournot equilibria relative to $(w_t)_{t \geq 0}$.

Proof: Theorem 1 yields continuous functions $\bar{w}(\theta)$ and $\tilde{y}(\theta)$ for $\theta \in K$. Define

$$\bar{w} := \inf\{\bar{w}(\theta) \mid \theta \in K\} > 0 \quad \text{and} \quad \tilde{y} := \sup\{\tilde{y}(\theta) | \theta \in K\} < y \quad .$$

Given $(w_t)_{t \geq 0}$ and $((y_t, p_t, np_t, y_t))_{t \geq 0}$ as in Corollary 1, Theorem 1 gives the existence of $\psi_t \in C(\mathbb{R}_{++})$ such that $(y_t, p_t, np_t y_t) \in \text{TSCE}(w, \psi_t)$ and $\psi_t(p_t) = p_{t+1}$ for all $t \geq 0$. Choose then any $\Psi_t \in C(\mathbb{R}^{t+1}_{++}, \mathbb{R}_{++})$ so that $\Psi_t(., p_{t-1}, \ldots, p_0) = \psi_t(.)$. □

Corollary 2: Let $\sigma(1) \neq 1$. Then there exist $\bar{w} > 0$ and $\tilde{y} < \bar{y}$ such that each (y, p, npy) with $\tilde{y} \leq y \leq \bar{y}$ and $p = \dfrac{M}{ny(1 - c(1))}$ is a stationary symmetric Cournot equilibrium relative to every $w \in [0, \bar{w}]$.

Since demand tends to infinity if the price tends to zero[5], it follows that in the perfectly competitive case where firms are price takers and the

[5]This holds irrespective of the forecast function, because of demand of the old generation.

price is determined by goods market equilibrium, unemployment does not prevail at arbitrary low wage rates. Therefore, one would expect that at low enough wage rates the interval of possible equilibrium allocations converges to the full employment allocation if the number of firms tends to infinity. The result is contained in the following theorem.

Theorem 2: Let $\hat{L} < \bar{L}$ and $\theta > 0$. There exist $N > 0$ and $\bar{w}(n) > 0$, $n > N$, so that for all $n > N$, $0 \leq w \leq \bar{w}(n)$, $\psi \in C(\mathbb{R}_{++})$, and $(y, p, I) \in \text{TSCE}(w, \psi)$ with $\psi(p) = \theta p$ it follows $nl(y) > \hat{L}$. Furthermore, N and $\bar{w}(n)$ are continuous in θ.

Proof: Assume $(y, p, I) \in \text{TSCE}(w, \psi)$ for some ψ with $\psi(p) = \theta p$. Then the income is determined by $I = M/(1-c(\theta))$. Define $d := (1+c(\theta))^{-1}+1$ and observe that

$$\tfrac{M}{Y} < P_\psi(Y, I) < d\tfrac{M}{Y} \qquad \forall\, Y > 0. \tag{4.5}$$

Set

$$\begin{aligned} N &:= \bar{L}l(l^{-1}(\bar{L} - \hat{L})d^{-1})^{-1} \quad \text{and} \\ \bar{w}(n) &:= \frac{M(l^{-1}(\bar{L} - \hat{L}) - dl^{-1}(\bar{L}/n))}{n\bar{L}l^{-1}(\bar{L}/n)} > 0 \quad , \; n > N. \end{aligned}$$

Let $n > N$ and $w \leq \bar{w}(n)$, and assume that $nl(y) =: L \leq \hat{L}$. From $(y, p, I) \in \text{TSCE}(w, \psi)$ follows that no firm gains from deviating to the full employment level, i.e. that

$$\begin{aligned} P_\psi((n-1)y + \hat{y}(y), I)\hat{y}(y) - P_\psi(ny, I)y &\leq w(\bar{L} - (n-1)l(y)) - wl(y) \\ &= w(\bar{L} - L) \quad . \end{aligned} \tag{4.6}$$

Both inequalities in (4.5) together with (4.6) yield

$$\frac{n\hat{y}(y)}{(n-1)y + \hat{y}(y)} < \tfrac{nw}{M}(\bar{L} - L) + d \quad .$$

Using the monotonicity and concavity of l^{-1}, the left hand side can be estimated to below against $l^{-1}(\bar{L} - L)/l^{-1}(\bar{L}/n)$, so that it follows

$$l^{-1}(\bar{L} - L) < \left(\tfrac{nw}{M}(\bar{L} - L) + d\right)l^{-1}(\tfrac{\bar{L}}{n}) \quad .$$

But this implies

$$w > \frac{M(l^{-1}(\bar{L}-L) - dl^{-1}(\bar{L}/n))}{(\bar{L}-L)nl^{-1}(\bar{L}/n)} \geq \bar{w}(n) \quad ,$$

a contradiction. □

The following two corollaries apply this result in order to show that unemployment becomes small in all perfect foresight equilibria and stationary equilibria if the number of firms is large and wages are low. In the case of general perfect foresight equilibria, it has to be assumed that inflation factors are bounded in order to obtain the bounds N and $\bar{w}(n)$. The proofs are straightforward.

Corollary 3: Let $\hat{L} < \bar{L}$ and let $K \subset \mathbb{R}_{++}$ be compact. Then there exist $N > 0$ and $\bar{w}(n)$, $n > N$, so that given $n > N$ and $(w_t)_{t \geq 0}$ with $0 \leq w_t \leq \bar{w}(n)$ $\forall t \geq 0$, all perfect foresight symmetric Cournot equilibria relative to $(w_t)_{t \geq 0}$ with inflation factors contained in K have higher employment than $\hat{L}$ in every period.

Corollary 4: Let $\hat{L} < \bar{L}$. Then there exist $N > 0$ and $\bar{w}(n)$, $n > N$, so that given $n > N$ and $0 \leq w \leq \bar{w}(n)$, all stationary symmetric Cournot equilibria relative to w have higher employment than $\hat{L}$.

4.3.4 A CES example

Consider now the special case where u is a utility function with constant elasticity of substitution $0 \leq \sigma \neq 1$ and discount factor equal to one, which implies $c(\theta) = (1 + \theta^{1-\sigma})^{-1}$. Furthermore, marginal costs are constant and normalized to 1, i.e. $l(y) = y$. The class of forecast functions is of the type

$$\Psi_t(p_t, \ldots, p_0) = p_t^{1+\beta} \Phi_t(p_{t-1}, \ldots, p_0)$$

with some $\beta \in \mathbb{R}$ and a continuous function Φ_t. This class contains log-linear adaptive price or inflation factor expectations

$$p^e_{t+1} = p_t^{1+\beta}(p^e_t)^{-\beta} \quad , -1 < \beta \leq 0, \tag{4.7}$$
$$\frac{p^e_{t+1}}{p_t} = \left(\frac{p_t}{p_{t-1}}\right)^{\beta}\left(\frac{p^e_t}{p_{t-1}}\right)^{1-\beta} \quad , 0 < \beta \leq 1, \tag{4.8}$$

and constant memory autonomous forecast rules of the type

$$\Psi_t(p_t, \ldots, p_0) = p_t^{1+\beta}\Phi(p_{t-1}, \ldots, p_{t-\tau}) \quad , t \geq \tau. \tag{4.9}$$

The following proposition shows that under this specification there exists a continuum of stationary Cournot equilibria with unemployment at arbitrarily low wage rates. Due to many simplifying estimations in the proof the assumptions are much stronger than needed.

Proposition 1: Assume $n \leq 14$. Then there exists a $\tilde{y} < \bar{y}$ such that each $(y, \frac{2M}{ny}, 2M)$ with $y \in [\tilde{y}, \bar{y}]$ is a stationary symmetric Cournot equilibrium relative to every $0 \leq w \leq (n-1)2M/(n\bar{L})$ for some forecast function with constant elasticity $1+\beta$ and $\beta \in \mathbb{R}$.

Before proving Proposition 1, the following lemma gives conditions guaranteeing sufficiency of the first order condition (4.4) for a temporary equilibrium, and it holds even under the general assumptions of Section 4.3.2.

Lemma 1: Let $p > 0$, $y := \dfrac{M}{np(1 - c(\psi(p)/p))}$, and $I := npy$. Assume

(i) D_ψ is continuously differentiable and $D'_\psi(.,I) < 0$,

(ii) $MR(p') := p'\left(1 + \dfrac{D_\psi(p',I)}{np'D'_\psi(p',I)}\right)$ is strictly monotonically increasing for $p_h \geq p' \geq p_l$ with $D_\psi(p_h, I) = (n-1)y$ and $D_\psi(p_l, I) = (n-1)y + \hat{y}(y)$,

(iii) (y, p) satisfies (4.4).

Then $(y, p, I) \in \mathrm{TSCE}(w, \psi)$.

Proof: The profit function of a single firm

$$\pi(y') := P_\psi((n-1)y + y', I)y' - wl(y')$$

is differentiable by (i). Using (ii) and (iii) it can be easily shown that $\pi'(y') > 0$ if $0 \leq y' < y$ and that $\pi'(y') < 0$ if $y < y' \leq \hat{y}(y)$. Therefore, y is the unique maximum of π on $[0, \hat{y}(y)]$. □

Proof of Proposition 1: Define $\tilde{y} := 2 \cdot 3^{-\frac{3+\sqrt{3}}{4+2\sqrt{3}}} \bar{y} < \bar{y}$, choose any $y \in [\tilde{y}, \bar{y}]$ and define $p := \frac{2M}{ny}$. Since $\sigma \neq 1$ define $\beta := 4(1 - \frac{p}{n(p-w)})/(\sigma - 1)$. (y, p, npy) is a stationary symmetric Cournot equilibrium relative to w if the assumptions of Lemma 1 are satisfied for

$$D_\psi(p', 2M) = \frac{M}{p'}\left(1 + 2c(p'^\beta p^{-\beta})\right) \quad ,$$

since the forecast function is $\psi(p') = p'^{1+\beta} p^{-\beta}$. (iii) is fulfilled with equality by definition of β.

To check condition (i), use (4.2) to show that $D'_\psi(., 2M) < 0$ if and only if

$$-1 + 2c((1-c)(\sigma - 1)\beta - 1) < 0 \qquad \forall\, c \in \ (0, 1).$$

But this can be shown to hold if $0 < (\sigma - 1)\beta < 2 + \sqrt{3}$ which follows from the assumptions on n and w.

To show condition (ii), observe that $MR(.)$ is strictly monotonically increasing if the elasticity of $D_\psi(., 2M)$ is monotonically decreasing. Using (4.2) again and the observation that $c(p'^\beta p^{-\beta})$ is monotonically increasing in p' since $(\sigma - 1)\beta > 0$, $D'_\psi(p', 2M)p'/D_\psi(p', 2M)$ is monotonically decreasing in p' if and only if $2c(1-c)(\sigma - 1)\beta(1 + 2c)^{-1}$ is monotonically decreasing in $c = c(p'^\beta p^{-\beta})$. But this can be shown to hold only for $c > (\sqrt{3} - 1)/2$, which implies that the elasticity of $D_\psi(., 2M)$ decreases monotonically for

$$p' > pc^{-1}((\sqrt{3} - 1)/2)^{1/\beta} = p(\sqrt{3})^{1/(\beta(1-\sigma))} =: \tilde{p} \quad .$$

A straightforward computation shows that

$$\begin{aligned} D_\psi(\tilde{p}, 2M) &= ny(\sqrt{3})^{1+\frac{1}{\beta(\sigma-1)}}/2 \\ &\geq ny3^{\frac{3+\sqrt{3}}{4+2\sqrt{3}}}/2 \geq n\bar{y} \quad , \end{aligned}$$

which implies $\tilde{p} \leq p_l$. Therefore, $MR(.)$ is strictly monotonically increasing on $[p_l, \infty)$. □

As an illustration of Proposition 1, Figure 4.2 shows a numerical example of the stationary equilibrium $(y, p, I) = (1, 1, 2)$ with unemployment for $n = 2$, $w = 0$, $\sigma = 3$, $\beta = 1$ and $M = 1$. The labor capacity constraint $\bar{L} > 2$ may not be too large, since marginal revenue becomes positive at large employment levels.

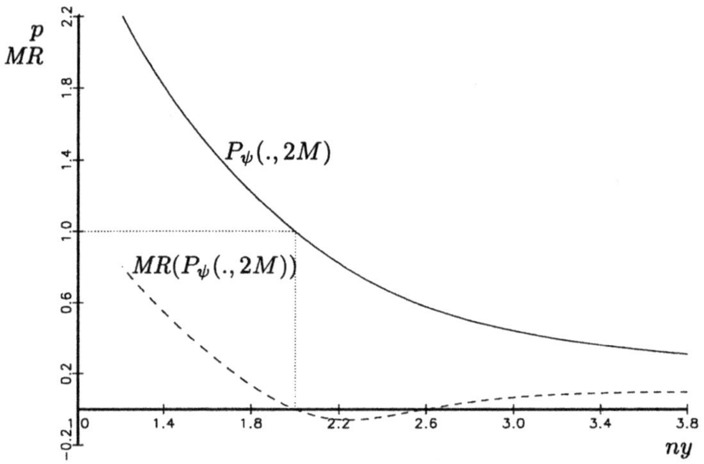

Figure 4.2: Stationary equilibrium with unemployment at $w = 0$.

It is important to notice that the existence of stationary equilibria with full employment and unemployment depends crucially on the expectations elasticity β. Consider again the first order condition (4.4) at a stationary equilibrium $(\frac{2M}{np}, p, 2M)$:

$$1 - \tfrac{1}{4}(\sigma - 1)\beta \geq \frac{p}{n(p-w)} \quad \text{and} \quad \frac{2M}{p} \leq \bar{L} \quad \text{(one with equality).}$$

Figure 4.3 illustrates the solutions p of this condition in dependence on β for a given $\sigma > 1$ and with the definition $\beta^* := 4(1-\frac{1}{n})/(\sigma-1)$. The sets of stationary equilibria differ qualitatively for $w = 0$ and for $w > 0$. While for $w = 0$ and $\beta = \beta^*$ there is a continuum of candidates of stationary unemployment equilibria, for $\beta < \beta^*$ only full employment equilibria emerge. For $w > 0$ there are unique stationary equilibria at different values of $\beta < \beta^*$ with unemployment if β is close to β^*. The interval of expectations elasticities β with unemployment converges to β^* if $w \to 0$. If w is larger than $2M/\bar{L}$, there are only unemployment equilibria for all values of $\beta < \beta^*$. For $\beta > \beta^*$ there exists no stationary equilibrium. One can show in this case that the first order condition (4.4) allows for solutions with a constant inflation and employment converging to zero. Such "stagflationary" equilibria have been found in another model by D'Aspremont, Dos Santos Ferreira & Gérard-Varet (1991).

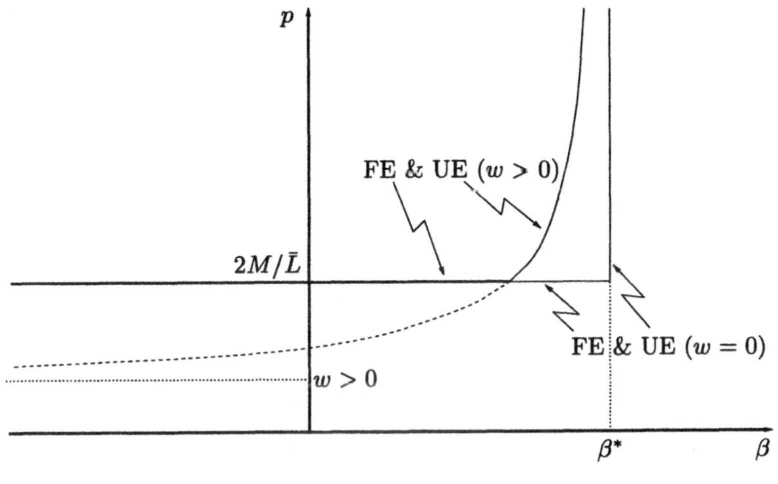

Figure 4.3: Stationary full employment (FE) and unemployment (UE) equilibrium prices p in dependence on the expectations elasticity β for $w = 0$ and $w > 0$.

In a model with multiple perfect foresight equilibria, it is reasonable to carry out an equilibrium selection with regard to some learning process.

In the model considered here, it is impossible to study such a learning process globally, since it is hard to verify existence of temporary equilibria at a given forecast rule for each parameter specification, and an explicit iteration formula cannot be computed. Instead, the so-called *local temporary equilibrium dynamics* (see Grandmont (1998)), which describes sequences of temporary equilibria close to a stationary equilibrium under some specific forecast rule, can be analyzed and it can be checked if stationary equilibria are stable or not. Such processes will be described only for the constant elasticity learning rules defined above, even though an endogenous determination of β (via OLS learning, for example) would be more relevant, since β plays the decisive rule for the occurrence of unemployment in a stationary equilibrium.

The temporary equilibrium dynamics is defined differently for unemployment and for full employment equilibria. In case of a stationary unemployment equilibrium (y, p, I) consider the implicit equation in (p_t, Φ_t)

$$\begin{aligned} c(p_t^\beta \Phi_t)\Big(1 - c(p_t^\beta \Phi_t)\Big)(\sigma - 1)\beta &= 1 - \frac{p_t}{n(p_t - w)} , \qquad (4.10) \\ \text{with} \quad ny_t = \frac{M}{p_t(1 - c(p_t^\beta \Phi_t))} &< \bar{L} , \end{aligned}$$

where $\Phi_t = (p_t^e)^{-\beta}$ in case of (4.7), $\Phi_t = (p_t^e)^{1-\beta} p_{t-1}^{-1}$ in case of (4.8), and $\Phi_t = \Phi(p_{t-1}, \ldots, p_{t-\tau})$ in case of (4.9). The full employment temporary equilibrium dynamics is described by the implicit equation [6]

$$\begin{aligned} \frac{M}{p_t(1 - c(p_t^\beta \Phi_t))} &= \bar{L} , \qquad (4.11) \\ \text{with} \quad c(p_t^\beta \Phi_t)\Big(1 - c(p_t^\beta \Phi_t)\Big)(\sigma - 1)\beta &< 1 - \frac{p_t}{n(p_t - w)} . \end{aligned}$$

$p_t^e, p_{t-1}, \ldots, p_{t-\tau}$ are assumed to be close to p, so that Φ_t is close to $p^{-\beta}$. If the assumptions of Lemma 1 are fulfilled for $\psi(p') = p'^{\beta+1} p^{-\beta}$, they are fulfilled also for $\psi(p') = p'^{\beta+1} \Phi_t$ with Φ_t close to $p^{-\beta}$ which implies that possible implicit solutions of (4.10) or (4.11) close enough to p yield in fact temporary symmetric Cournot equilibria.

[6] The boundary case of two binding constraints in a stationary equilibrium will not be considered.

Proposition 2: Stationary unemployment equilibria relative to $w > 0$ are stable under (4.7), (4.8) and (4.9). Stationary unemployment equilibria relative to $w = 0$ are stable under (4.7) and (4.8), and are stable under (4.9) if and only if the corresponding price is stable under $p_t = \Phi(p_{t-1}, \ldots, p_{t-\tau})^{-1/\beta}$. Stationary full employment equilibria are stable under (4.7), and may be unstable under (4.8) and (4.9).

Proof: Consider first the case of a stationary unemployment equilibrium price p relative to $w > 0$. By the implicit function theorem (4.10) has a solution $p_t = \Gamma(\Phi_t)$ with $\Gamma'(p^{-\beta}) = 0$. This implies that p is locally stable under (4.7), (4.8) and (4.9).

If p is a stationary unemployment equilibrium price at $w = 0$, $\beta = 4(1 - 1/n)/(\sigma - 1)$ is independent of p and there is a continuum of stationary equilibria at β. (4.10) is then

$$c(p_t^\beta \Phi_t)(1 - c(p_t^\beta \Phi_t)) = 4 \quad ,$$

which has the solution $p_t = \Phi_t^{-1/\beta}$. Under (4.7) it follows $p_{t+1}^e = p_t = p_t^e$ so that the initially expected price level determines the stationary equilibrium. Under (4.8) it follows $p_{t+1}^e = p_t = (p_t^e)^{1-1/\beta} p_{t-1}^{1/\beta}$ so that the initial price and the initial expected price determine the stationary equilibrium.

If p is a stationary full employment equilibrium price, the implicit function theorem yields a solution $p_t = \Gamma(\Phi_t)$ of (4.11) if $\beta(\sigma - 1) \neq 2$, and it is $\Gamma'(p^{-\beta}) = p^{1+\beta} \frac{\sigma-1}{2-\beta(\sigma-1)}$. For (4.7), it is

$$\left. \frac{dp_{t+1}^e}{dp_t^e} \right|_{p_t^e = p} = -\frac{\beta(\sigma + 1)}{2 - \beta(\sigma - 1)} \in [0, 1)$$

since $\beta \in (-1, 0]$. In case of (4.8) and (4.9) the local dynamics can be unstable since $\Gamma'(p^{-\beta})$ can be arbitrary large for $\beta(\sigma - 1)$ close to 2. For example, in case of (4.8) the local dynamics is described by $q_{t+1} = \Gamma(q_t)^{-\beta^2} q_t^{1-\beta}$ with $q_t = (p_t^e)^{1-\beta)} p_{t-1}^{-1}$. The derivative at the steady state $q = p^{-\beta}$ is $1 - \frac{2\beta}{2-\beta(\sigma-1)}$ and may be larger than 1 or less than -1. □

Notice that such a strong stability result for unemployment equilibria does not need to hold if some assumptions are relaxed. Already a discount factor different from one or a non-linear technology would imply $\Gamma'(p^{-\beta}) \neq 0$ generally, so that stability can no longer be guaranteed. In this respect, Proposition 2 should only be understood as an example showing that for one plausible model specification simple learning mechanisms lead to stable unemployment equilibria, while full employment equilibria may be unstable. But still it is very special and in particular it does not endogenize the determination of the crucial forecast elasticity parameter.

4.4 Indeterminacy of rational expectations equilibria

4.4.1 Are there unique rational forecasts?

One might suspect that the above indeterminacy of PFE and stationary equilibria rests on the arbitrariness of forecast function away from the equilibrium states, and that, as in Rankin (1997) and in Example 2 of Section 4.2, equilibria are determinate if forecast functions are rational. However, these models use the artificial assumption of separated output markets for young and for old consumers, implying that the equilibrium price in the market for old persons is independent of the behavior and thereby the expectations of young persons.

By use of two examples, it will be shown in this section that the result of Rankin rests on this special assumption. If young and old persons consumed in the same market, the equilibrium price in period $t+1$ depends on price expectations of young persons for period $t+2$. As a consequence, rationality of the forecast function in period t depends on the forecast function in period $t+1$ whose rationality depends on the next forecast function, and so on. It turns out that this feedback of tomorrows forecasts on todays rational expectations leads again to indeterminacy of REE.

4.4.2 Characterization of rational expectations equilibria

To show indeterminacy of REE, a general result will be proven which characterizes REE by a simple geometric condition on sets of temporary equilibria. Formally, for $A \subset X$ let

$$U(A) := \left\{(x, x_+) \in A \times A \mid \exists\, \psi : X \to A \text{ with } x \in \mathrm{T}(\psi) \text{ and } \psi(x) = x_+\right\}.$$

Theorem 3: Let $A \subset X$ and let $g : A \to A$ fulfill $\mathrm{Graph}(g) \subset U(A)$. Then each trajectory of g, i.e. each sequence $(x_t^*)_{t \geq 0}$ with $x_0^* \in A$ and $x_{t+1}^* = g(x_t^*)$ for all $t \geq 0$, is a rational expectations equilibrium.

Proof: Let $x \in A$. Since $(x, g(x)) \in U(A)$, there exists $\psi(.; x) : X \to A$ such that $\psi(x; x) = g(x)$ and $x \in \mathrm{T}(\psi(.; x))$.

Let $(x_t^*)_{t \geq 0}$ be a trajectory of g. Define $(\Psi_t)_{t \geq 0}$ recursively as follows:

$$\Psi_0(x_0) := \psi(x_0; x_0^*) \quad , \; x_0 \in X \, ,$$

and for $t > 0$

$$\Psi_t(x_t, \ldots, x_0) := \psi(x_t; \Psi_{t-1}(x_{t-1}, \ldots, x_0)) \quad , \; (x_t, \ldots, x_0) \in X^{t+1} \, .$$

It has to be shown that $(\Psi_t)_{t \geq 0}$ is rational and that $(x_t^*)_{t \geq 0}$ is a PFE generated by $(\Psi_t)_{t \geq 0}$.

$(\Psi_t)_{t \geq 0}$ is rational: Let $\tau \geq 0$ and $(x_\tau, \ldots, x_0) \in X^{\tau+1}$. Define $(x_t)_{t > \tau}$ recursively by $x_{t+1} = \Psi_t(x_t, \ldots, x_0)$ for all $t \geq \tau$. For $t > \tau$ it follows by construction

$$\begin{aligned} x_t &\in \mathrm{T}\Big(\psi(.;x_t)\Big) \\ &= \mathrm{T}\Big(\psi(.;\Psi_{t-1}(x_{t-1},\ldots,x_0))\Big) \\ &= \mathrm{T}\Big(\Psi_t(.,x_{t-1},\ldots,x_0)\Big) \quad . \end{aligned}$$

$(x_t^*)_{t\geq 0}$ is a PFE, i.e. $x_{t+1}^* = \Psi_t(x_t^*,\ldots,x_0^*)$ and $x_t^* \in \mathrm{T}(\Psi_t(.,x_{t-1}^*,\ldots,x_0^*)$ for all $t \geq 0$: For $t = 0$ it is, by definition of $\psi(.;.)$, $x_1^* = g(x_0^*) = \psi(x_0^*;x_0^*) = \Psi_0(x_0^*)$ and $x_0^* \in \mathrm{T}(\psi(.;x_0^*)) = \mathrm{T}(\Psi_0(.))$. For $t > 0$ it is

$$\begin{aligned} x_{t+1}^* &= g(x_t^*) = \psi(x_t^*;x_t^*) \\ &= \psi(x_t^*;\Psi_{t-1}(x_{t-1}^*,\ldots,x_0^*)) \\ &= \Psi_t(x_t^*,\ldots,x_0^*) \quad . \end{aligned}$$

Rationality of $(\Psi_t)_{t\geq 0}$ finally yields $x_t^* \in \mathrm{T}(\Psi_t(.,x_{t-1}^*,\ldots,x_0^*))$ for all $t > 0$. □

This theorem will be applied in the following subsections to show the indeterminacy of REE in two examples. The general method is to find appropriate subsets $A \subset X$ such that $U(A)$ is a "thick" set containing the graphs of many functions $g : A \to A$. Indeterminacy of stationary REE follows if there is a (non–trivial) subset A such that $\mathrm{dia}(A) := \{\ (a,a) \mid a \in A\} \subset U(A)$, since then, by the above theorem, each $a \in A$ is a stationary REE.

4.4.3 Example 1: Intertemporal Cournot–Walras equilibria

Consider an economy which is almost identical to the one in Section 4.3.2. There are again three goods (labor, consumption good and money), representative two–period living overlapping generations, and $n \geq 2$ identical firms producing a single consumption good from labor with labor requirement function l which fulfills Assumption 2. The difference to Hart's model is that now firms do not take the wage rate as given. Instead, they perceive correctly the adjustment of wages to their labor market clearing level *after* they choose their production plans. Thus, they are Cournot oligopolists both on the output and on the labor market, a type of competition which has been called "oligempory" and which has recently been investigated in a different model by Gaygisiz & Madden (1997). Firms also correctly perceive the income feedback on demand. Therefore, the equilibrium to be considered here is a temporary equilibrium of the Cournot–Walras type as introduced in Section 3.2.3.

Assume that there is disutility of labor and that preferences of a representative young household are represented by a utility function of the type

$$u(c_1, c_2) - v(L) \quad . \tag{4.12}$$

Assumption 3: $u : \mathbb{R}^2_+ \rightarrow \mathbb{R}$ and $v : \mathbb{R}_+ \rightarrow \mathbb{R}$ are C^1 and strictly monotonic. u is strictly quasi–concave and linearly homogeneous, and $u'_i(c_1, c_2) \rightarrow \infty$ if $c_i \rightarrow 0,\ c_j > 0,\ j \neq i$, for $i = 1, 2$. v is convex.

Such types of utility functions have been used, for instance, by Hart (1982), Section 5, and by Dixon & Rankin (1994). The additive separability combined with linear homogeneity rules out any wealth effect on labor supply.

Let $\pi \geq 0$ denote the profit income of the young household, w the wage rate, p the actual goods price and p^e the expected price for the next period. The solution of the decision problem

$$\max\ (4.12) \quad \text{s.t.} \quad pc_1 + m \leq wL + \pi \quad , 0 \leq L \quad \text{and} \quad c_1, m \geq 0,$$
$$p^e c_2 \leq m \quad , \ c_2 \geq 0 \ .$$

fulfills

$$c_1 \ = \ c(\frac{p^e}{p})\frac{wL + \pi}{p} \ ,$$

$$\begin{aligned} c_2 &= (1 - c(\frac{p^e}{p}))\frac{wL + \pi}{p^e} , \\ v'(L) &= \frac{w}{\mathbf{P}(p, p^e)} \quad \text{if } 0 < L , \end{aligned}$$

where $0 < c(.) < 1$ is the propensity to consume as defined in 4.3.2, and where

$$\mathbf{P}(p, p^e) := u\Big(\frac{c(p^e/p)}{p}, \frac{1 - c(p^e/p)}{p^e}\Big)^{-1}$$

is the intertemporal price index. Denote by $\psi : \mathbb{R}_+ \to \mathbb{R}_+$ a price forecast function, i.e. $p^e = \psi(p)$, and by $M > 0$ the money stock held by the old generation.

If Y is aggregate output and L aggregate employment, aggregate profit income is $\pi = pY - wL$ and the goods market equilibrium identity $c(\frac{\psi(p)}{p})Y + \frac{M}{p} = Y$ can be solved for

$$Y = \frac{M}{p(1 - c(\psi(p)/p))} \quad . \tag{4.13}$$

Assume that c and ψ are such that this identity can be inverted, and write $p = P_\psi(Y)$ for the inverse demand function.

For each $L \geq 0$ the labor market clearing wage is $w = v'(L)\mathbf{P}(p, p^e)$, and write

$$W_\psi(Y, L) := v'(L)\mathbf{P}\Big(P_\psi(Y), \psi(P_\psi(Y))\Big)$$

for the inverse labor supply function. Define the profit function of a single firm if all others produce y by

$$\pi_\psi(y', y) := y'P_\psi\Big((n-1)y+y'\Big) - W_\psi\Big((n-1)y+y', (n-1)l(y)+l(y')\Big)l(y') \quad .$$

Definition 7: Let $\psi \in C(\mathbb{R}_{++})$. A *temporary symmetric Cournot–Walras equilibrium relative to* ψ is a tuple (y, p, w) such that

$$\begin{aligned} &\text{(i)} \quad y \in \operatorname*{argmax}_{y' \geq 0} \pi_\psi(y', y) \quad , \\ &\text{(ii)} \quad p = P_\psi(ny) \quad , \\ &\text{(iii)} \quad w = W_\psi(ny, nl(y)) \quad . \end{aligned}$$

The set of output prices in temporary symmetric Cournot–Walras equilibria relative to ψ is denoted $\mathrm{T}(\psi)$.[7]

Notice that the implicit exclusion of inefficient production plans, i.e. not to consider production plans (y, ℓ) with $\ell > l(y)$, is no restriction, since a reduction of the labor input does not increase the wage rate (because of convexity of v) and lowers therefore costs without changing revenue.

A *rational expectations Cournot–Walras equilibrium* is defined as a REE in 4.2 for the state space $\mathbb{R}_{++}$ and the above temporary equilibrium map $\mathrm{T}(\psi)$, and a *stationary Cournot–Walras equilibrium* is a stationary REE.

Consider for the rest of this subsection the following specification of the economy: u is Cobb–Douglas with discount factor $(1-\alpha)/\alpha$, i.e.

$$u(c_1, c_2) = (\frac{c_1}{\alpha})^{\alpha}(\frac{c_2}{1-\alpha})^{1-\alpha}$$

such that

$$c(.) \equiv \alpha \quad \text{and} \quad \mathbf{P}(p, p^e) = p^{\alpha}(p^e)^{1-\alpha} \quad .$$

Furthermore, assume

$$\begin{aligned} v(L) &= \frac{a}{\nu+1}L^{\nu+1} \quad , \nu \geq 0, a > 0, \\ l(y) &= k \cdot y \quad , k > 0 \,. \end{aligned}$$

This implies that the inverse output demand and labor supply functions are

$$P_{\psi}(Y) = P(Y) := \frac{M}{Y(1-\alpha)} \quad , \; W_{\psi}(Y, L) = aL^{\nu}P(Y)^{\alpha}(\psi(P(Y))^{1-\alpha} \,.$$

It is quite useful to consider first perfectly competitive equilibria where firms take the price and wage as given. That is, a *temporary perfectly competitive equilibrium* (y, p, w) is a solution of $p = kw = P(ny)$ and

$$w = a(nky)^{\nu}p^{\alpha}\psi(p)^{1-\alpha} \,.$$

[7] Notice that the output price p contains all relevant information: Given p, there is at most one (y, w) such that (y, p, w) is a temporary symmetric Cournot–Walras equilibrium.

Given a future output price p_+, it can be easily shown that there exists a unique temporary perfectly competitive equilibrium price p such that $\psi(p) = p_+$. This price is given by

$$p = p^c(p_+) := \left(ak^{1+\nu} p_+^{1-\alpha} (\frac{M}{1-\alpha})^\nu\right)^{1/(1-\alpha+\nu)} .$$

Therefore, the function p^c describes the backward perfect foresight dynamics under perfect competition. It is easy to verify that it has a unique fixpoint $\bar{p}^c$.

In contrast to the perfectly competitive case, rational expectations Cournot–Walras equilibria and also stationary Cournot–Walras equilibria are highly indeterminate. This will follow from the following theorem which shows that the sets $U(A)$ are large, provided that A is an unbounded interval:

Theorem 4: Let $p_0 > 0$ and let $A := [p_0, \infty)$. Then

$$\left\{(p, p_+) \in A^2 \mid p \geq p^c(p_+)\right\} \subset U(A) .$$

The proof of this theorem needs the following two lemmata.

Lemma 2: Let $(p, p_+) \in \mathbb{R}_{++}$ such that $p \geq p^c(p_+)$. Then there exists a continuous $\phi : \mathbb{R}_{++} \to \mathbb{R}_{++}$ such that $p \in \mathrm{T}(\phi)$ and $p_+ = \phi(p)$.

Proof: Let $\phi(P) := \mu P^\gamma$ where $\mu := p_+ p^{-\gamma}$, which implies $p_+ = \phi(p)$. $\gamma \in \mathbb{R}$ will be chosen in the following such that $p \in \mathrm{T}(\phi)$.

A straightforward computation yields

$$\pi_\phi(y', y) = \frac{M}{1-\alpha}\left(\frac{y'}{y' + (n-1)y} - B\frac{y'}{(y' + (n-1)y)^\lambda}\right)$$

with

$$B := ak^{\nu+1}\mu^{1-\alpha}\left(\frac{M}{1-\alpha}\right)^{\alpha+\gamma(1-\alpha)-1} \text{ and } \lambda := \alpha + \gamma(1-\alpha) - \nu .$$

Clearly, y is the output of a single firm in a temporary symmetric Cournot–Walras equilibrium if it is a solution of the fixpoint problem

$$y \in \operatorname*{argmax}_{y' \geq 0} \pi_\phi(y', y) \quad . \tag{4.14}$$

This solution of this problem can be computed as

$$y \begin{cases} = & \frac{1}{n}\left(\frac{B(n-\lambda)}{n-1}\right)^{1/(\lambda-1)} & \text{if} \quad \lambda < 1 \, , \\ \in & [0,\infty) \text{ arbitrary} & \text{if} \quad \lambda = 1, \; B = 1 \, . \end{cases} \tag{4.15}$$

The corresponding temporary equilibrium price is then $\dfrac{M}{ny(1-\alpha)}$.

Assume first that $p > p^c(p_+)$, and choose γ such that

$$\lambda := \alpha + \gamma(1-\alpha) - \nu \stackrel{!}{=} n - (n-1)\left(\frac{p}{p^c(p_+)}\right)^{1-\alpha+\nu} \quad . \tag{4.16}$$

Notice that $\lambda < 1$ because of $p > p^c(p_+)$. Thus, from (4.15), the temporary equilibrium price is

$$\frac{M}{ny(1-\alpha)} = \frac{M}{1-\alpha}\left(\frac{B(n-\lambda)}{n-1}\right)^{1/(1-\lambda)} \, ,$$

which equals p because of (4.16) and the definitions of B and p^c. Hence, $p \in \mathrm{T}(\phi)$.

Consider next the case $p = p^c(p_+)$ and choose γ such that $\lambda = 1$. But now $p = p^c(p_+)$ implies $B = 1$ and it follows from (4.15) that $y = \dfrac{M}{np(1-\alpha)}$ solves (4.14). Thus, $p \in \mathrm{T}(\phi)$. □

Lemma 3: Let $\psi, \phi : \mathbb{R}_{++} \to \mathbb{R}_{++}$ and let $p > 0$ such that $p \in \mathrm{T}(\phi)$, $\psi(p) = \phi(p)$ and $\psi \geq \phi$. Then, $p \in \mathrm{T}(\psi)$.

Proof: Observe that $p \in \mathrm{T}(\phi)$ if and only if $y := \dfrac{M}{np(1-\alpha)}$ solves (4.14).

From $\psi \geq \phi$, it follows $\pi_\psi \leq \pi_\phi$, and $\phi(p) = \psi(p)$ yields $\pi_\phi(y,y) = \pi_\psi(y,y)$. Therefore, y also solves (4.14) with π_ϕ replaced by π_ψ. Hence, $p \in \mathrm{T}(\psi)$. □

Proof of Theorem 4: Let $(p, p_+) \in A^2$ such that $p \geq p^c(p_+)$. It has to be shown that there exists a continuous $\psi : \mathbb{R}_{++} \to [p_0, \infty)$ such that $p \in \mathrm{T}(\psi)$ and $p_+ = \psi(p)$. Take ϕ from Lemma 2 and define $\psi(P) := \max(\phi(P), p_+)$. Thus, $\psi(p) = p_+$ and $\psi \geq \phi$. Lemma 3 implies that $p \in \mathrm{T}(\psi)$. □

Corollary 5: Let $g : [\bar{p}^c, \infty) \to [\bar{p}^c, \infty)$ fulfill $g(p) \leq (p^c)^{-1}(p)$ for all $p \geq \bar{p}^c$. Then all trajectories of g are rational expectations Cournot–Walras equilibria. In particular, each $\bar{p} \geq \bar{p}^c$ is a stationary Cournot–Walras equilibrium.

Proof: From the assumption, it follows $p \geq p^c(g(p))$ for all $p \geq \bar{p}^c$. Theorem 4 implies $\mathrm{Graph}(g) \subset U([\bar{p}^c, \infty))$ and Theorem 3 yields the first claim. Since $p^c(p) \leq p$ for all $p \geq \bar{p}^c$, it follows $\mathrm{dia}([\bar{p}^c, \infty)) \subset U([\bar{p}^c, \infty))$. Therefore, each $\bar{p} \geq \bar{p}^c$ is a stationary Cournot–Walras equilibrium. □

Figure 4.4 illustrates $p^c(.)$, $\bar{p}^c$, and one arbitrary function g satisfying the requirement of Corollary 5.

There are two important remarks: First, a surprising implication of this result is that there is no convergence of rational expectations Cournot–Walras equilibria to perfectly competitive REE if the number of firms tends to infinity: The set of REE obtained in Corollary 5 is independent of the number of firms! This is in contrast to the corollaries following Theorem 2 for Hart's model. However, for a REE $(p_t)_{t \geq 0}$ for which $p_t > p^c(p_{t+1})$ in one period, one can see from (4.16) in the proof of Lemma 2 that λ and thereby the elasticity of the forecast function γ tend to $-\infty$ as n tends to infinity. Therefore, steeper and steeper forecast functions are needed in order to support $(p_t)_{t \geq 0}$ as a REE. One would expect that convergence to perfectly competitive REE takes place if one imposes boundedness restrictions on the slopes of forecast functions.

Second, one could suspect that the indeterminacy of REE is a consequence of the profit maximization hypothesis as it is the case for inde-

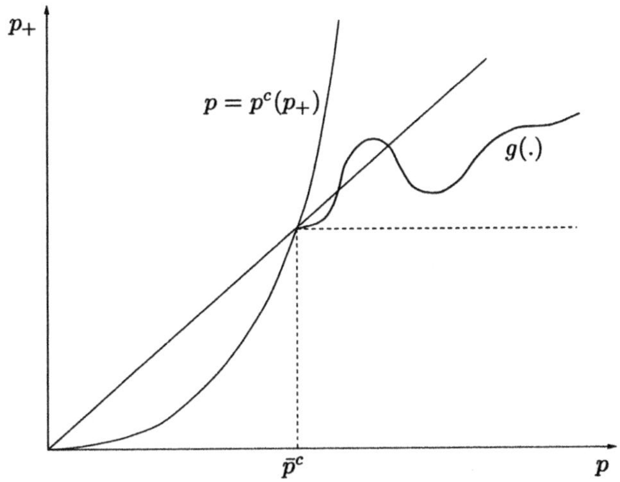

Figure 4.4: An arbitrary function g generating rational expectations Cournot–Walras equilibria.

terminacies resulting from different choices of price normalization (see Section 3.2.3), and that this indeterminacy vanishes if shareholders maximized their indirect utilities. However, this is not true. Assume that the single houschold is replaced by $n+1$ households, one worker and n shareholders, each owning exactly one firm, such that the worker has the same preferences as above over consumption and leisure, and such that the shareholders have the same preferences u over consumption streams. It follows then from linear homogeneity of u and separability of the worker's preferences that the inverse demand functions P_ψ and W_ψ are the same as before. Replace now the profit maximization hypothesis in (i) of Definition 7 by the maximization of each shareholder's indirect utility which is

$$U(y', y) := \frac{\pi_\psi(y', y)}{\mathbf{P}\Big(P_\psi(y' + (n-1)y), \psi(P_\psi(y' + (n-1)y))\Big)} .$$

One can show easily with a modified proof of Lemma 2 that Theorem 4 remains valid. The next subsection illustrates a similar indeterminacy result in a pure exchange economy.

4.4.4 Example 2: A market game in an OLG exchange economy

Consider the following pure exchange overlapping generations model. There is a single non–storable consumption commodity and fiat money. In each period $t = 0, 1, \ldots$ there are born n two–period living consumers. Each consumer has an endowment $\omega > 0$ of the consumption good when he is young, and zero endowment when he is old. Let $\Omega := n\omega$ be the aggregate endowment in each period. Each of the n old consumers in period 0 holds an amount $m > 0$ of fiat money. Let $M := nm$ be the aggregate money endowment. Each consumer has the same utility function $u : \mathbb{R}^2_+ \to \mathbb{R}$ representing preferences over consumption plans in their lifetime which is assumed to be strictly monotonic, strictly quasi–concave and continuous.

As proposed by Shapley & Shubik (1977), assume that in each period there is a trading post exchanging fiat money for the consumption good. Assume that we are in a certain period and suppress for the moment the time index. Suppose that old consumer $i \in I := \{1, \ldots, n\}$ in this period holds an amount m_i of money.

Each young consumer $i \in I$ places at the trading post an *offer* $x_i \in [0, \omega]$, each old consumer places a *bid* $\mu_i \in [0, m_i]$ of money. All offers and bids are placed simultaneously. Given all offers and bids, there is a market clearing rule defined in the following way. If $\mu := \sum_{i \in I} \mu_i$ and $X := \sum_{i \in I} x_i$ are both different from zero, there is trade at the market clearing price $p := \mu/X$. If X or μ is equal to zero, there is no trade. In this case, the young generation receives no money at all, and the economy becomes non–monetary and sticks to the no–trade equilibrium for all subsequent periods.

If $X > 0$, it is a dominant strategy of each old consumer to bid all money.

The old consumer i maximizes consumption and solves

$$\max_{0 \le \mu_i \le m_i} \frac{\mu_i}{\mu_i + \mu_{-i}} \cdot X \quad ,$$

which has the unique solution $\mu_i = m_i$ since $\mu_{-i} := \sum_{j \neq i} \mu_j$ must be strictly positive, if other consumers maximize their consumption, too. Thus, each old consumer bids all money he owns, which implies that the aggregate money stock is equal to M in all periods, as long as $X > 0$ in all periods.

Each young consumer, knowing that he bids all money in the next period, forms expectations about the price in the next period. Since aggregate bids of old consumers are M, this is equivalent to forming expectations about the aggregate supply X^e of young consumers in the next period. Assume that all young consumers expect that aggregate supply in the next period depends on aggregate supply X in the current period according to a continuous forecast function $\psi : [0, \Omega] \to [0, \Omega]$, i.e. $X^e = \psi(X)$. The expected market clearing price is therefore $p^e = M/\psi(X)$.

If $(x_i)_{i \in I}$ is a list of offers, consumption of young consumer j in the current period is $\omega - x_j$ and expected consumption in the next period is

$$\frac{px_j}{p^e} = \begin{cases} \frac{x_j}{\sum_{i \in I} x_i} \psi(\sum_{i \in I} x_i) & \text{if } \sum_{i \in I} x_i > 0 \, , \\ 0 & \text{if } \sum_{i \in I} x_i = 0 \, . \end{cases}$$

Definition 8: Let $\psi \in C([0, \Omega])$. $X \in [0, \Omega]$ is a *temporary symmetric trading post equilibrium relative to* ψ, either if

$$0 < \frac{X}{n} \in \operatorname*{argmax}_{0 \le \xi \le \omega} u\Big(\omega - \xi, \frac{\xi}{\xi + (n-1)\frac{X}{n}} \psi(\xi + (n-1)\tfrac{X}{n})\Big)$$

or if

$$0 = X \in \operatorname*{argmax}_{0 \le \xi \le \omega} u(\omega - \xi, \psi(\xi)) \quad .$$

The set of temporary symmetric trading post equilibria relative to ψ is $\mathrm{T}(\psi)$.

This definition cannot be immediately extended to a sequence of temporary equilibria as in Section 4.2. If $X = 0$ is a temporary equilibria in one period, the economy becomes non–monetary and all temporary equilibria in subsequent periods must be no–trade equilibria. Thus, the above definition can only describe sequences of temporary equilibria, as long as $X > 0$ in all periods. However, under the following assumption on u and ψ, a no–trade temporary equilibrium can be excluded.

Assumption: u is continuously differentiable on $\mathbb{R}_{++}$ and $\lim_{x\to 0} u_2'(\omega, x) = \infty$, and ψ is differentiable with $\psi'(0) > 0$.

Under this assumption, $0 \notin \mathrm{T}(\psi)$, since for small enough $\varepsilon > 0$ it follows $u(\omega - \varepsilon, \psi(\varepsilon)) > u(\omega, 0)$. A *rational expectations trading post equilibrium* is defined as a REE in 4.2 for the state space $[0, \Omega]$ and the above temporary equilibrium map $\mathrm{T}(\psi)$, and a *stationary trading post equilibrium* is a stationary REE.

Theorem 5: Let $\Omega \geq C > 0$ and $A = (0, C)$. Then

$$\Big\{(nx, nx_+) \in A^2 \mid u(\omega - x, x_+) > u(\omega, 0)\Big\} \subset U(A) \quad .$$

Proof: Let $0 < x, x_+ < c := C/n$ and $u(\omega - x, x_+) > u(\omega, 0)$. One has to show that there exists $\psi : [0, \Omega] \to (0, C)$ such that $\psi(nx) = nx_+$ and $nx \in \mathrm{T}(\psi)$.

Choose a continuous function $\Phi : [0, \omega] \to [0, c)$ such that

$$\Phi(x) = x_+ \ ,$$

$$u(\omega - \xi, \Phi(\xi)) < u(\omega - x, x_+) \ \forall\, \xi \neq x \text{ and } \Phi(\xi) < \xi \tfrac{c}{x} \ \forall\, \xi > 0 \ , \quad (4.17)$$

$$\Phi \text{ is differentiable in } 0 \text{ and } \tfrac{c}{x} > \Phi'(0) > 0 \ .$$

Such a Φ exists as Figure 4.5 illustrates.

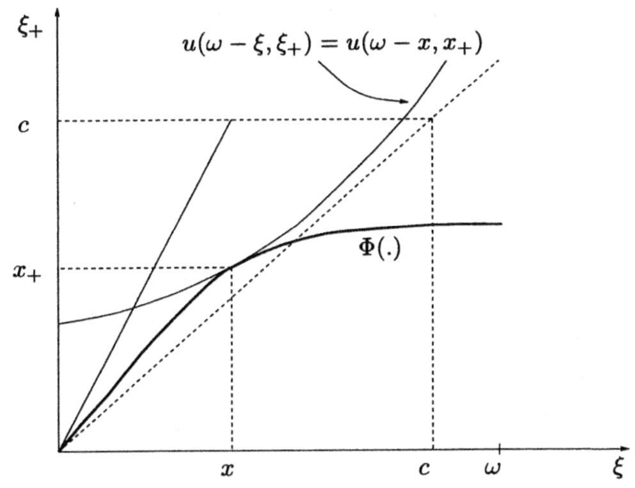

Figure 4.5: A function Φ fulfilling (4.17).

First, define $\psi(X)$ for $(n-1)x < X \leq (n-1)x+\omega$ as follows:

$$\psi(X) := \frac{\Phi(X-(n-1)x)X}{X-(n-1)x} \quad .$$

It follows $\psi(nx) = nx_+$ and $\psi(X) > 0$. If $X \leq nx$, $\Phi(\xi) < \frac{c}{x}\xi$ implies $\psi(X) < \frac{cX}{x} \leq C$. If $X > nx$, $\Phi(.) < c$ implies $\psi(X) < \dfrac{CX}{n(X-(n-1)x)} < C$. By definition of Φ, ψ can be continuously extended in $X = (n-1)x$ by $\psi((n-1)x) = \Phi'(0)(n-1)x \in (0,C)$. Therefore, $\psi(X) \in (0,C)$ for all $X \in [(n-1)x,(n-1)x+\omega]$.

Second, extend ψ for $X \notin [(n-1)x,(n-1)x+\omega]$ arbitrary and continuous such that $0 < \psi(.) < C$.

Finally, if a consumer deviates to $\xi \in [0,\omega]$, $x \neq \xi$, he gets worse off since

$$u\Big(\omega-\xi,\frac{\xi}{\xi+(n-1)x}\psi(\xi+(n-1)x)\Big) = u(\omega-\xi,\Phi(\xi)) < u(\omega-x,x_+) \quad .$$

Therefore, $nx \in \mathrm{T}(\psi)$. □

The following corollary shows that all individually rational trajectories are REE:

Corollary 6: Let $\bar{x} > 0$ be chosen such that either $\bar{x} < \omega$ and $u(\omega - \bar{x}, \bar{x}) = u(\omega, 0)$ or $\bar{x} = \omega$ and $u(0, \omega) \geq u(\omega, 0)$. Let $g : (0, \bar{x}) \to (0, \bar{x})$ fulfill $u(\omega - x, g(x)) > u(\omega, 0)$ for all $0 < x < \bar{x}$. Then for each trajectory $(x_t^*)_{t \geq 0}$ of g, $(nx_t^*)_{t \geq 0}$ is a rational expectations trading post equilibrium. Furthermore, each $X \in (0, n\bar{x})$ is a stationary trading post equilibrium.

Proof: Define $G(X) := ng(X/n)$. Theorem 5 implies Graph$(G) \subset U((0, n\bar{x}))$. Theorem 3 proves the first claim. Furthermore, $0 < X < n\bar{x}$ implies $u(\omega - \frac{X}{n}, \frac{X}{n}) > u(\omega, 0)$ and thus dia$(0, n\bar{x}) \subset U((0, n\bar{x}))$. This gives the second claim. □

Figure 4.6 illustrates the set of individual rational allocations and one arbitrary function g generating REE.

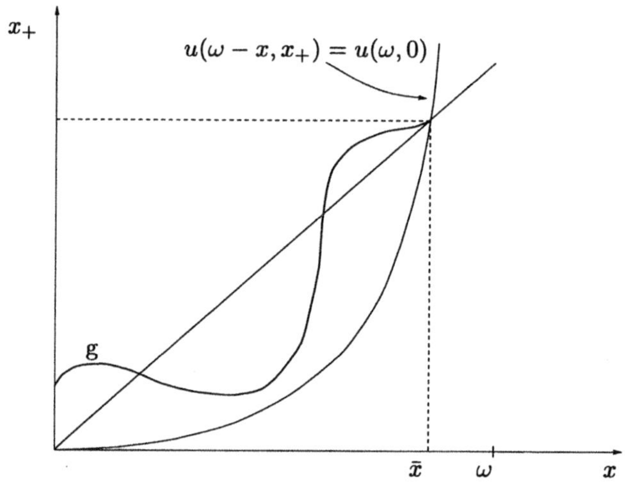

Figure 4.6: An arbitrary function g generating rational expectations trading post equilibria.

As in the previous section, there is no convergence of rational expectations trading post equilibria to perfectly competitive REE as n tends to infinity, since allocations of rational expectations equilibria are independent of n. Again, in the proof of Theorem 5, this may be attributed to the possibility of arbitrarily steep forecast functions. If the function Φ is differentiable in x, the forecast function is differentiable in nx with slope

$$\psi'(nx) = \frac{x_+}{x} + n(\Phi'(x) - \frac{x_+}{x}) \quad .$$

Observe that $\Phi'(x)$ is equal to the marginal rate of substitution (see Figure 4.5):

$$\Phi'(x) = \frac{u_1'}{u_2'}(\omega - x, x_+) \quad .$$

Further, the offer curve, which describes price–taking behavior, is described by

$$\frac{x_+}{x} = \frac{u_1'}{u_2'}(\omega - x, x_+) \quad .$$

Therefore, if the pair (x, x_+) is not on the offer curve, the slope of the forecast functions supporting x as a temporary equilibrium and forecasting x_+ diverges if $n \to \infty$.

4.5 Summary and discussion

In this chapter, macroeconomic models with imperfect competition have been considered in a temporary equilibrium framework. Temporary equilibrium models are characterized by static equilibrium concepts applied to a single period of an economy in which agents' expectations about future periods depend on current and past variables. The equilibrium concepts to be used in almost all of this literature is that of an objective equilibrium, since it is regarded by most researchers as superior to subjective and therefore more arbitrary equilibrium concepts.

To avoid arbitrariness of temporary equilibria resulting from the choice of arbitrary forecast functions, it is natural to characterize those forecast functions which are rational or which are at least consistent with perfect foresight along one equilibrium path. The resulting sequences of temporary equilibria with perfect foresight may therefore be regarded as free of any "irrationality" in the economy, since no agent makes any forecast errors. In a REE it is even the case that no agent would make forecast errors if there were deviations from equilibrium states in the past. Thus, one would expect that such PFE or REE, if they exist at all, are highly determinate in the sense that intertemporal states are linked by some non–degenerate equation, or that they can be generated by a backward or forward perfect foresight dynamics.

However, quite the opposite occurs! It turns out that these equilibria are highly indeterminate. There exist almost arbitrary REE, as long as they fulfill certain conditions, and there exists typically also a continuum of stationary equilibria. In Hart's model this leads to the possibility of stationary (perfect foresight) equilibria with involuntary unemployment. In OLG models with Cournot competition between firms on the labor and output market, and in a market game in an OLG exchange economy, there is even a continuum of stationary equilibria which do not converge to the perfectly competitive stationary equilibria if the number of strategic agents tends to infinity.

The explanation of this phenomenon is a kind of *folk theorem* argument.[8] Different states in each period may be supported as a temporary equilibrium by some forecast function. But this forecast function is rational, if the states that are forecasted for the next period can itself be generated as temporary equilibria by the forecast functions of the next period, and so on. Of course, if the economy ended in some finite period, the whole argument would break down.

This analysis should not give the impression that existence of temporary equilibria with objective demand is unproblematic as soon as forecast functions are rational. Instead, the results show that sequences of forecast functions may be constructed for which temporary equilibria exist *and* which are rational. However, given some forecast functions, existence of

[8]In contrast to the folk theorem, this result is based on the variation of fundamentals of the economy, namely expectation functions. In this respect, the result is conceptually more related to the Sonnenschein–Debreu–Mantel result on the arbitrariness of excess demand functions.

a temporary equilibrium is at least as problematic as it is in static general equilibrium models. Moreover, equilibrium selection via learning cannot be adequately carried out in this setting, since for a given learning process described by a sequence of forecast functions, a corresponding sequence of temporary equilibria may not exist, and therefore a globally defined dynamical system cannot be analyzed.[9]

[9]This is of course also the case in perfectly competitive models where often only an implicit temporary equilibrium dynamics with learning is described, see Grandmont (1998).

Chapter 5

A dynamic macroeconomic model with price setting firms

"*In the last case* (of decreasing returns) *there will be an indeterminate tract through which the index of value will oscillate, or rather will vibrate irregularly for an indefinite length of time.*"

F.Y. Edgeworth (1897)

"*Under conditions of disequilibrium, there is no reason that there should be a single market price, and we may very well expect that each firm will charge a different price. (...) the whole adjustment process is apt to be very irregular. Although the broad tendency will be for prices to rise when demand exceed supply, there can easily be a considerable dispersion of prices among different sellers of the same commodity, as well as considerable variability over time in the rate of change of prices.*"

K.J. Arrow (1959)

5.1 Introduction

The perfectly rational approach of the previous chapter (firms know the objective demand curve and households make perfect price forecasts) turned out not to determine unique intertemporal equilibria. To select among such equilibria by learning, it is not very promising to maintain the assumption of a temporary objective equilibrium in each period, and thereby to suppose perfectly rational firms. This is for two reasons. First, as shown in 4.3.4, there is a continuum of stationary equilibria and it depends crucially on the learning process of price expectations which equilibrium is attained. Thus, "learning" cannot be expected to provide a true selection of equilibria. Second, temporary objective equilibria may not exist, even if the usual assumptions on preferences, technologies and expectation functions are imposed which are sufficient for existence of temporary competitive equilibria. Thus, a globally defined dynamics cannot be expected to exist, and even if it exist, it need not be computable.

In order to study a globally defined dynamical system, it seems therefore necessary to avoid the objective demand assumption and to consider a model where also the learning process of imperfectly competitive agents, as formulated in a general setting in Chapter 2, takes place in "real time". Thus, this chapter focuses on intertemporal macroeconomics from a strictly dynamic point of view in the sense that no temporary equilibrium concept will be used in the definition of the dynamic process. [1]

Another goal of this chapter is to formulate a model where the determination of prices is not left with an invisible hand but is a result of decisions of price setting firms. This will be done in a model whose basic ingredients are very similar to the model in Section 4.3. There are again only two markets, one for exchanging labor and one for exchanging consumption goods against money. There are n imperfectly competitive

[1] A previous version of this chapter was published in Kaas (1997).

firms producing a homogeneous output good from labor, and households are again characterized by an overlapping generations structure.

As in Hart's model, there is also a sequential market structure, but with a strict dynamic interpretation. Exchange on the labor market takes place before the goods market opens and therefore also before goods demand may be observed by firms. This implies that labor demand depends on the output demand expectations of firms which have to be specified explicitly. Dynamic models with such a sequential market structure and competitive price adjustment have been analyzed by Böhm (1998) and Weddepohl (1996). In this chapter an imperfectly competitive goods market will be considered. It is assumed that firms form expectations about one shift parameter of a subjective demand function, which corresponds to the IOF in the framework of Chapter 2. The household faces uncertainty concerning this parameter and estimates it with Bayesian updating, or he expects the parameter with certainty and estimates it with adaptive expectations or other learning rules. Similar ideas of price adjustment processes under imperfect competition in related models have been discussed already in the seventies by Iwai (1974) and Benassy (1977), but they have never been analyzed rigorously.

Assume that wage rates are fixed for the duration of a period and are adjusted by a competitive wage adjustment process from period to period. Thus, this chapter will provide a dynamic price adjustment model of an economy which is almost identical to the one in 4.3 and it can again be investigated whether unemployment can occur under competitive wage and imperfectly competitive price adjustment.

The chapter is organized as follows. The model in a general non-parameterized form is introduced in 5.2. Section 5.3 describes a parameterization of the economy and of firms demand expectations, which allows an explicit computation of the firms' price and production decisions and which is similar to the parameterization of Iwai (1974). This parameterized version is analyzed in Section 5.4, first for the case of a single monopolistic firm, and then for the case of many firms.

5.2 The economy

The economy evolves in discrete time periods $t = 1, 2, \ldots$. There are n firms, a household sector described by two-period living overlapping generations, and a government. In each period, there are three types of commodities: labor, a consumption good and money. Each firm produces the consumption commodity from labor as the only input factor, takes the wage rate as given and decides on the output and price to maximize the expected profit. It is assumed here that households consume only in the second period of their life. Only young households supply labor and receive wage and profit income which they save as money for retirement.[2]

Markets open sequentially in the following way. In the first stage the labor market opens with a given wage rate. Firms determine their labor demands and households decide on labor supply. Actual trade is given by the minimum of aggregate demand and supply. In the second stage the goods market opens. Firms supply their output and announce their goods prices, to be fixed during the period. Consumers and the government enter the market and direct their demand to the lowest price firms. Trade occurs again at the minimum of supply and demand with higher price firms getting the residual demand in case of demand rationing. Firms may store possible excess supply for the next period.

[2]It is not difficult to extend the model for consumption of young households, but one has to be careful with income payments: Because of the sequential market structure, the first period consumption has to be financed from labor income alone, since profits are distributed only at the end of the period.

5.2.1 The agents

The economic agents are characterized in detail by the following assumptions:

The households:

Households live for two periods. Each young household supplies labor and receives net wage and profit income which he saves as money for consumption in the second period of his life. Preferences over labor supply and future consumption $(L, c) \in [0, L_{\max}] \times \mathbb{R}_+$ are represented by a continuous utility function of the form
$u(L_{\max} - L, c)$ which is strictly quasi-concave and strictly monotonic in both arguments.

A young household who faces the wage rate w_t and expects profit income at the end of the period π_t^e and a commodity price p_{t+1}^e for the next period, solves the intertemporal utility maximization problem

$$\max u(L_{\max} - L_t, c_{t+1}) \qquad \text{s.t.}$$

$$0 \le L_t \le L_{\max} \quad \text{and} \quad p_{t+1}^e c_{t+1} \le w_t L_t + \pi_t^e \quad .$$

The optimal labor supply is a function $L^s(w_t, p_{t+1}^e, \pi_t^e)$ which is homogeneous of degree zero. Price and profit expectations are assumed to be described by forecast functions depending only on current information.

An old household, who holds the amount M_t of money at the beginning of period t and faces k different prices $p_t^1 < \ldots < p_t^k$ and demand constraints $\bar{c}_t^1, \ldots, \bar{c}_t^k$, trades $c_t^l = \min(\tilde{c}_t^l, \bar{c}_t^l)$, $l = 1, \ldots, k$, where the effective demands are

$$\begin{aligned} \tilde{c}_t^1 &= \frac{M_t}{p_t^1} \quad , \\ \tilde{c}_t^l &= \max\left(0, \frac{M_t - \sum_{m=1}^{l-1} p_t^m c_t^m}{p_t^l}\right) \; , \quad 1 < l \le k \; . \end{aligned}$$

It is furthermore assumed that possible forced savings of an old household are transferred to the young generation.

The firms:

Each firm produces the output commodity from labor via a production function $F : \mathbb{R}_+ \to \mathbb{R}_+$. (For the moment, firms indices are suppressed.) It is assumed that the firm faces a subjective demand function $y_t(p_t) = Q_t D(p_t)$, where the function D is monotonically decreasing and subjectively given, and $Q_t > 0$ is a shift parameter. The firm believes that Q_t is a random variable with distribution function Φ_t. If Φ_t is a Dirac distribution concentrated at q_t, Q_t is identified with q_t. Firms maximize current expected profits which are distributed at the end of the period to the young generation.

Given a stock of possibly past unsold goods supply z_t and the wage rate w_t, the expected profit is a function depending on the price p and employment L defined by

$$E_t\Pi(p, L) := \int_0^\infty p \min(qD(p), F(L) + z_t)\Phi_t(q)dq - w_t L \quad . \tag{5.1}$$

The firm, facing a demand constraint on the labor market $\bar{L}$, solves

$$\max_{p,L} \quad (5.1) \qquad \text{s.t.} \qquad L \leq \bar{L} \quad . \tag{5.2}$$

Assume that the solution of the price decision problem at a given supply $x \geq 0$ exists and is given by the function

$$p^*(x, \Phi_t) := \operatorname*{argmax}_p \int_0^\infty p \min(qD(p), x)\Phi_t(q)dq \quad .$$

Assume furthermore that $E_t\Pi(p^*(F(.) + z, \Phi_t), .)$ is a concave function in L.[3] If the unconstrained optimal employment level is $L^*(\Phi_t, w_t, z_t)$, the solution of (5.2) is

$$L_t = \min(L^*(\Phi_t, w_t, z_t), \bar{L}) \quad \text{and} \quad p_t = p^*(F(L_t) + z_t, \Phi_t) \quad . \tag{5.3}$$

[3]These assumptions will be fulfilled for the parameterization of D, Φ_t and F in the next section.

$L_t^d := L^*(\Phi_t, w_t, z_t)$ is the firm's labor demand and $x_t^s := F(L_t) + z_t$ its goods supply. If y_t is the demand addressed to the firm, its sales are $x_t = \min(x_t^s, y_t)$ and the profit/loss is $p_t x_t - w_t L_t$.

After the firm observes demand y_t at the price p_t, it learns the "true" shift parameter $q_t^* := y_t / D(p_t)$. Using this information, the firm forms expectations about the distribution of the shift parameter of the next period according to a learning rule $\mathcal{L}$:

$$\Phi_{t+1} = \mathcal{L}(\Phi_t, q_t^*) \quad .$$

The government:

The government levies a proportional tax rate $\tau \in [0,1]$ on wage and profit income and demands in each period an amount $g \geq 0$ of the consumption good. Deficits resp. surpluses are financed via money creation resp. destruction. It is assumed that the government minimizes expenditures so that at prices $p_t^1 < \ldots < p_t^k$ and demand constraints $\bar{g}_t^1, \ldots, \bar{g}_t^k$ it trades $g_t^l = \min(\tilde{g}_t^l, \bar{g}_t^l)$, $l = 1, \ldots, k$, where effective demands are

$$\begin{aligned} \tilde{g}_t^1 &= g \ , \\ \tilde{g}_t^l &= \max\left(0, g - \sum_{m=1}^{l-1} g_t^m\right) \ , \ 1 < l \leq k \ . \end{aligned}$$

5.2.2 The markets

After expressing their demands and supplies on each market, some agents are rationed in case of disequilibrium. For simplification, a proportional rationing scheme on each market is employed, where it is assumed that agents do not perceive their influence on rationing constraints. Therefore, the rationing scheme is (subjectively) non-manipulable (cf. Böhm (1989)). Since demands and constraints on the markets are determined sequentially, this approach does not use any fixed-price equilibrium concept to make demands and market signals consistent.

The labor market:

Labor supply is equal to $L_t^s = L^s(w_t, p_{t+1}^e, \pi_t^e)$, while labor demands are given by $L_{t,i}^d = L_i^*(\Phi_{t,i}, w_t, z_{t,i})$, $i = 1, \ldots, n$. This leads to labor demand constraints

$$\bar{L}_{t,i} = L_{t,i}^d \cdot \frac{L_t^s}{\sum_{j=1}^n L_{t,j}^d} \quad , \; i = 1, \ldots, n \; .$$

Aggregate labor demand is $L_t^d = \sum_{j=1}^n L_{t,j}^d$ and total employment is $L_t = \min(L_t^d, L_t^s)$. To the next period, the wage rate adjusts competitively dependent on the disequilibrium on the labor market:

$$w_{t+1} = w_t(1 + \mathcal{W}(L_t^s, L_t^d)) \quad , \tag{5.4}$$

where the wage inflation factor $\mathcal{W}$ is a function depending on labor supply and demand and fulfills (see e.g. Böhm, Lohmann & Lorenz (1994))

$$\begin{aligned} \mathcal{W}(L_t^s, L_t^d) &\geq 0 \quad \text{if} \quad L_t^d \geq L_t^s \quad , \\ \mathcal{W}(L_t^s, L_t^d) &\leq 0 \quad \text{if} \quad L_t^d \leq L_t^s \quad . \end{aligned}$$

The goods market:

Firms enter the goods market with their supplies $x_{t,i}^s$ and announce prices $p_{t,i}$. Even though there is one homogeneous consumption good, firms can charge different prices. Consumers observe these prices and visit the firms sequentially, starting at the lowest price. Suppose that firms set $k \leq n$ different prices $p_t^1 < \ldots < p_t^k$. Let $I_t^l \subset \{1, \ldots, n\}$ be the set of firms charging price p_t^l. Thus, $p_t^l = p_{t,i}$ if and only if $i \in I_t^l$. Each $l = 1, \ldots, k$ is regarded as a particular market with aggregate supply and demand given by

$$X_t^l := \sum_{i \in I_t^l} x_{t,i}^s \qquad \text{and} \qquad Y_t^l := \tilde{c}_t^l + \tilde{g}_t^l \quad .$$

The proportional rationing rule implies that demand and supply constraints are

$$\bar{c}_t^l = \tilde{c}_t^l \frac{X_t^l}{Y_t^l} \; , \quad \bar{g}_t^l = \tilde{g}_t^l \frac{X_t^l}{Y_t^l} \; , \; \text{and} \quad y_{t,i} = x_{t,i}^s \frac{Y_t^l}{X_t^l} \quad \text{for} \quad i \in I_t^l \quad .$$

This yields transactions

$$c_t^l = \min(\bar{c}_t^l, \tilde{c}_t^l) \quad , \quad g_t^l = \min(\bar{g}_t^l, \tilde{g}_t^l)$$

$$\text{and} \quad x_{t,i} = \min(y_{t,i}, x_{t,i}^s) \quad \text{for} \quad i \in I_t^l \quad .$$

In accordance with the theory of temporary equilibria with quantity rationing (see e.g. Böhm (1989)), allocations may be distinguished with regard to their rationing situation. Since on the goods market trade takes place at many firms with different prices, one may define that there is excess demand (supply resp.) if and only if at firms with the highest price there is excess demand (supply resp.), i.e. if and only if $Y_t^k > X_t^k$ ('$<$' resp.). Define the following subsets of the state space:

K (Keynesian Unemployment): Excess supply on the goods and on the labor market.

C (Classical Unemployment): Excess supply on the labor market and excess demand on the goods market.

I (Repressed Inflation): Excess demand on the goods and on the labor market.

U (Underconsumption): Excess demand on the labor market and excess supply on the goods market.

Furthermore, there are five boundary cases, one of them is the case of market clearing on both markets (**M**).

Obviously, this classification of disequilibrium regimes does not coincide with the one obtained under price taking behavior, since demands and supplies are defined differently. In particular, the market clearing state differs from the competitive equilibrium of the economy. It must be stressed, however, that some authors use the price-taking disequilibrium regimes also in a price setting framework, even though they do not really describe disequilibria on markets (cf. Benassy (1977)).

5.2.3 The dynamics

The above economy can be formulated as a dynamical system with strategic interaction in the framework of Chapter 2. The outcome in period t is a vector $(w_t, M_t, (z_{t,i})_i, (y_{t,i})_i)$, while an action of firm i in period t is $(x^s_{t,i}, p_{t,i})$. Only the demands $(y_{t,i})_i$ are influenced by the actual action choices, while the other parts of the outcome, w_t, M_t and $(z_{t,i})_i$, are predetermined by the previous period's data.

The definition of the dynamical system is as follows. Given M_t, w_t, p^e_{t+1}, π^e_t and $(\Phi_{t,i}, z_{t,i})^n_{i=1}$, demands, supplies and transactions on the labor and goods markets are defined as above. The young household saves net wage and profit income and receives possible forced savings of the old generation such that money savings for the next period are

$$\begin{aligned} M_{t+1} &= (1-\tau)\Big(w_t L_t + \sum_{i=1}^{n}(p_{t,i}x_{t,i} - w_t L_{t,i})\Big) + \Big(M_t - \sum_{l=1}^{k} p^l c^l_t\Big) \\ &= (1-\tau)\Big(\sum_{i=1}^{n} p_{t,i}x_{t,i}\Big) + \Big(M_t - \sum_{l=1}^{k} p^l c^l_t\Big) \quad . \end{aligned}$$

Using the feasibility of transactions $c^l_t + g^l_t = \sum_{i \in I^l_t} x_{t,i}$, this identity may be equivalently expressed as the government budget constraint

$$M_{t+1} - M_t = \sum_{l=1}^{k} p^l_t g^l_t - \tau\Big(\sum_{i=1}^{n} p_{t,i}x_{t,i}\Big) \quad . \tag{5.5}$$

Unsold goods are stored and depreciate at a common rate $\delta \in [0,1]$:

$$z_{t+1,i} = (1-\delta)(x^s_{t,i} - x_{t,i}) \quad , \; i = 1, \dots, n \; . \tag{5.6}$$

Expectations about the distribution of the shift parameter evolve according to

$$\Phi_{t+1,i} = \mathcal{L}_i(\Phi_{t,i}, y_{t,i}/D_i(p_{t,i})) \quad , \; i = 1, \dots, n \; . \tag{5.7}$$

Equations (5.5), (5.6), (5.7) and (5.4) together with a specification of price and profit expectations define the complete dynamical system. Figure 5.1 illustrates the time structure and the interaction between agents in this model. The encircled symbols denote markets for goods (X) and labor (L), while young (Y) and old (O) households, firms (F_i) and the government (G) are denoted by rectangles.

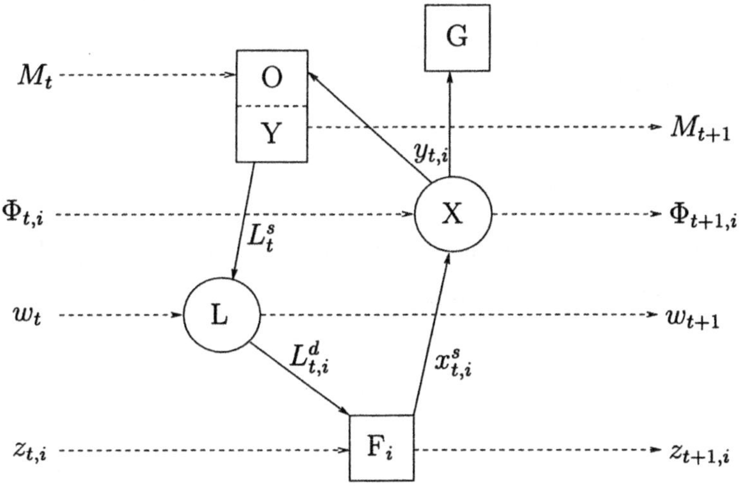

Figure 5.1: The time structure of the economy: dynamic variables (dashed lines) and supplies and demands (solid lines).

5.3 A parametric specification

5.3.1 Decision and expectation formation of firms

Consider a specific parameterization of the technology and of demand expectations of a typical firm $i \in \{1, \ldots, n\}$. In general, neither production technologies nor demand expectations need be identical for all firms. However, the index i will be suppressed as long as it is not needed.

Assume that the production function is

$$F(L) = \frac{A}{B} L^B \quad , \quad A, B > 0 \; .$$

The demand function D is a constant elasticity function $D(p) = p^{-\eta}$ with parameter $\eta > 1$. The decision problem and the expectation formation of the shift parameter will be considered separately for the case of certainty where the firm updates the parameter q_t with adaptive expectations, and for the case of uncertainty where the firm uses Bayesian updating for the estimation of Φ_t.

Adaptive expectations:

Assume that Φ_t is a Dirac distribution concentrated in q_t. Firms, observing ex–post the "true" shift parameter $q_t^* = y_t p_t^{\eta_i}$, use adaptive expectations to forecast the next shift parameter q_{t+1}. That is, the evolution of demand expectations (5.7) follows

$$q_{t+1,i} = b_i q_{t,i} + (1 - b_i) y_{t,i} p_{t,i}^{\eta_i} \quad , \; i = 1, \ldots, n \; , \tag{5.8}$$

with adjustment parameters $b_i \in (0, 1)$.

(5.2) can be written

$$\max_{p,L} p \cdot \min(q_t p^{-\eta}, F(L) + z_t) - w_t L \; .$$

Given the supply $x_t^s = F(L_t) + z_t$, the revenue maximizing price equates expected demand to supply, implying that the price setting rule is

$$p_t = p^*(x_t^s, q_t) := \left(\frac{q_t}{x_t^s}\right)^{1/\eta} \quad . \tag{5.9}$$

The expected profit function is therefore

$$\Pi(L) := q_t^{1/\eta} \left(\frac{A}{B} L^B + z_t\right)^{1-1/\eta} - w_t L \; .$$

Π is strictly concave and has a unique profit maximizing employment level L_t^d if $B \in (0, \frac{\eta}{\eta - 1})$. It can be computed analytically if there is no storage ($z_t = 0$),

$$L_t^d = \left(A^{1-1/\eta} B^{1/\eta} (1 - 1/\eta) q_t^{1/\eta} w_t^{-1}\right)^{\eta/(\eta(1-B)+B)} \quad , \tag{5.10}$$

or if there are constant returns ($B = 1$),

$$L_t^d = \max\left(0, \left(A^{1-1/\eta}(1-1/\eta)q_t^{1/\eta}w_t^{-1}\right)^\eta - \frac{z_t}{A}\right) \quad .$$

Bayesian updating:

The firm assumes that Q_t is a non–negative random variable which has a conditional exponential distribution

$$\Phi(q|\mu) := \frac{1}{\mu}e^{-q/\mu} \quad .$$

The non–negative parameters μ_t are uncertain and are estimated with Bayesian updating. Assume that prior beliefs about μ_t are inverted-gamma distributed, i.e. $1/\mu_t$ is gamma distributed, so that μ_t has the density function

$$\gamma_t(\mu) := G(\mu|\alpha_t, \beta_t) := \frac{\beta_t^{\alpha_t}}{\Gamma(\alpha_t)}\mu^{-(\alpha_t+1)}e^{-\beta_t/\mu}$$

with parameters $\alpha_t > 1$ and $\beta_t > 0$. This implies that the marginal probability density function of Q_t is a so–called gamma-gamma distributed (see e.g. Bernando & Smith (1993), p. 120):

$$\Phi_t(q) := \int_0^\infty \Phi(q|\mu)\gamma_t(\mu)d\mu = \alpha_t\beta_t^{\alpha_t}\frac{1}{(\beta_t+q)^{\alpha_t+1}} \quad .$$

The firm observes the ex–post position of the shift parameter $q_t^* = y_t p_t^\eta$, and computes the posterior density function of μ_t according to the Bayesian formula

$$\hat{\gamma}_t(.|q_t^*) = \frac{\Phi(q_t^*|.)\cdot\gamma_t(.)}{\Phi_t(q_t^*)} \quad .$$

Since gamma distributions are a conjugate family for exponentially distributed samples (see DeGroot (1970), p. 166), updated beliefs about μ_t after the observation of q_t^* are again inverted-gamma:

$$\hat{\gamma}_t(.|q_t^*) = G(.|\alpha_t+1, \beta_t+q_t^*) \quad .$$

The firm believes that (μ_s) follows a Markovian process $\mu_{s+1} = \lambda_s \mu_s$ with independently and identically distributed non-negative random variables λ_s which are assumed to be Pareto distributed with density function

$$\psi(\lambda) := \alpha b^{\alpha} \lambda^{-(\alpha+1)} \quad \text{for} \quad \lambda \geq b$$

with parameters $b > 0$ and $\alpha > 1$. Note that $E(\lambda_s) = \alpha b/(\alpha - 1)$, so that $b := (\alpha - 1)/\alpha$ implies that (μ_s) is believed to be a martingale, i.e. $E(\mu_s) = E(\mu_{s+1})$.

The prior probability density function of $\mu_{t+1} = \lambda_t \mu_t$ is computed by the formula

$$\gamma_{t+1}(\mu) = \int_0^{\infty} \hat{\gamma}_t(\mu/\lambda | q_t^*) \psi(\lambda)/\lambda \, d\lambda \quad . \tag{5.11}$$

An easy computation shows that γ_{t+1} is again an inverted-gamma distribution:

$$\gamma_{t+1}(.) = G(.|\frac{\alpha}{\alpha+1}(\alpha_t + 1), b(\beta_t + q_t^*)) \quad .$$

This implies that the updating process (5.7) can be expressed by the parameterization

$$\left(\alpha_t, \beta_t\right) \mapsto \left(\frac{\alpha}{\alpha+1}(\alpha_t + 1), b(\beta_t + q_t^*)\right) \quad .$$

Assuming that the initial parameter α_0 equals α implies that all α_t equal α and that the updating equations reduce to an updating of β_t alone. Since the mean value of Q_t is $q_t := E_t(Q_t) = \frac{\beta_t}{\alpha - 1}$, updating of β_t may be expressed equivalently by adaptive expectations of mean values:

$$q_{t+1,i} = b_i q_{t,i} + (1 - b_i) y_{t,i} p_{t,i}^{\eta_i} \quad , \; i = 1, \ldots, n \, . \tag{5.12}$$

Concerning the decision problem of the firm, consider first the price setting problem at employment level L_t and supply $x_t^s = F(L_t) + z_t$. This amounts to the maximization of expected revenue

$$\begin{aligned} E_t \mathrm{R}(p, x_t^s) &:= \int_0^{\infty} p \min(x_t^s, q p^{-\eta}) \Phi_t(q) dq \qquad (5.13) \\ &= \int_0^{x_t^s p^{\eta}} p^{1-\eta} q \Phi_t(q) dq + \int_{x_t^s p^{\eta}}^{\infty} p x_t^s \Phi_t(q) dq \end{aligned}$$

$$= p^{1-\eta}\Big(\Big(\frac{\beta_t}{\beta_t + x_t^s p^\eta}\Big)^\alpha \Big(\frac{\beta_t + \alpha x_t^s p^\eta}{1-\alpha}\Big) + \frac{\beta_t}{\alpha - 1}\Big)$$
$$+ p x_t^s \Big(\frac{\beta_t}{\beta_t + x_t^s p^\eta}\Big)^\alpha \quad .$$

One can show easily that

$$\lim_{p\to 0} E_t\mathrm{R}(p, x_t^s) = 0 \qquad \text{and} \qquad \lim_{p\to\infty} E_t\mathrm{R}(p, x_t^s) = 0 \quad ,$$

the latter because of $\eta > 1$. This implies that there exists a global maximum p^* for all x_t^s. A simple, but lengthy computation shows that the first order condition $\frac{\partial E_t\mathrm{R}}{\partial p}(p, x_t^s) = 0$ has the unique solution

$$p^*(x_t^s, q_t) := \Big(\frac{\gamma(\alpha,\eta) q_t}{x_t^s}\Big)^{1/\eta} \quad , \tag{5.14}$$

which determines the revenue maximizing price. The constant $\gamma(\alpha,\eta)$ is the unique positive solution of the equation

$$\Big(1 + \frac{\gamma}{\alpha - 1}\Big)^\alpha - 1 = \frac{(\alpha\eta - 1)\gamma}{(\eta - 1)(\alpha - 1)} \quad ,$$

and $q_t = \beta_t/(\alpha - 1)$ is the mean value of Q_t.

As a consequence of (5.14), firms typically do not set the price such that expected demand equals their supply. If and only if $\gamma(\alpha,\eta) < 1$, the firm chooses a price below the expected market clearing level such that expected demand $q_t p_t^{-\eta}$ is greater than supply x_t^s, and vice versa. Iwai (1974) has shown for exponentially distributed demand functions that this condition depends crucially on the demand elasticity. If the price elasticity of demand η is low enough, the firm sets its price so high that expected demand falls short of supply. The reason is that a price increase causes a weaker increase of the risk of being rationed if expected demand decreases slowlier. This forces firms to set the price the higher the less price–elastic the demand is.

Proposition 1: For any $\alpha > 1$ there exists $\hat{\eta}(\alpha) > 1$ such that $qp^*(x,q)^{-\eta} < x$ if and only if $\eta < \hat{\eta}(\alpha)$. The same statement holds for replacing '$<$' by '$>$' and '$=$'.

Proof: From the definition of p^*, $qp^*(x,q)^{-\eta} < x$ if and only if $\gamma(\alpha,\eta) > 1$. Since γ is monotonically decreasing in η, tends to 0 for $\eta \to \infty$, and tends to infinity for $\eta \to 1$, define $\hat{\eta}(\alpha) := \{\eta | \; \gamma(\alpha,\eta) = 1\}$ and the claim follows. □

Given the price setting function (5.14), maximal expected revenue (5.13) can be computed as

$$E_t\mathrm{R}(p^*(x,q_t),x) = c(\alpha,\eta)p^*(x,q_t)x \qquad \text{with}$$

$$c(\alpha,\eta) := \frac{1-(1+\gamma(\alpha,\eta)/(\alpha-1))^{1-\alpha}}{\gamma(\alpha,\eta)} < 1 \quad .$$

Therefore, expected profit at employment L is

$$E_t\Pi(p^*(F(L)+z_t,q_t),L) = c(\alpha,\eta)p^*(F(L)+z_t,q_t)(F(L)+z_t) - w_tL \quad ,$$

which is strictly concave in L and has a unique maximum $L_t^d \geq 0$ if F is not too convex. For instance, for the constant elasticity production function $F(L) = A/B \cdot L^B$ with parameters $A > 0$, $B \in (0, \frac{\eta}{\eta - 1})$, $E_t\Pi$ is concave in L and the first order condition for profit maximization is

$$L^{1-B}\Big(\frac{A}{B}L^B + z_t\Big)^{1/\eta} \leq Ac(\alpha,\eta)(\gamma(\alpha,\eta)q_t)^{1/\eta}(1-1/\eta)w_t^{-1}$$

with equality if $L > 0$. It allows analytic solutions if there is no storage ($z_t = 0$),

$$L_t^d = \Big(A^{1-1/\eta}B^{1/\eta}c(\alpha,\eta)(\gamma(\alpha,\eta)q_t)^{1/\eta}(1-1/\eta)w_t^{-1}\Big)^{\eta/(\eta(1-B)+B)} \quad , \tag{5.15}$$

or if returns are constant ($B = 1$),

$$L_t^d = \max\Big(0, \Big(A^{1-1/\eta}c(\alpha,\eta)(\gamma(\alpha,\eta)q_t)^{1/\eta}(1-1/\eta)w_t^{-1}\Big)^{\eta} - \frac{z_t}{A}\Big) \quad .$$

Figure 5.2 shows contour lines of the expected profit function, the expected demand curve, and also the price setting rule p^* for $\alpha = 2$, $\eta = 1.5 < \hat{\eta}(2) = 2$, which implies that the firm set its price higher than the level equating expected demand to supply.

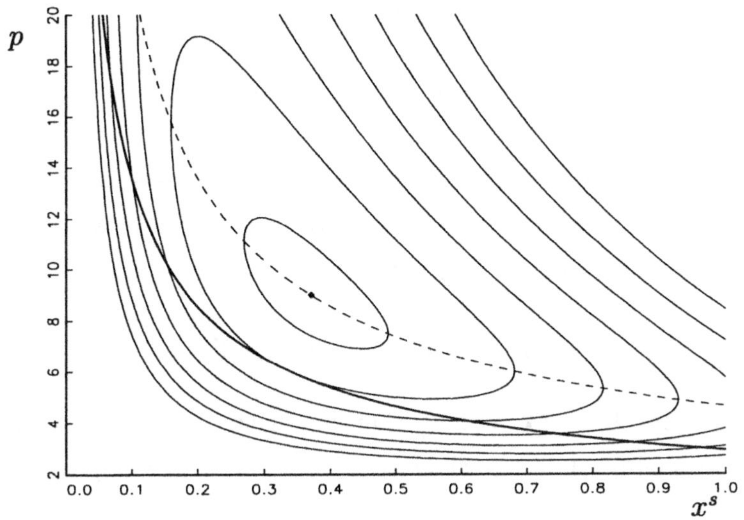

Figure 5.2: Contour lines of expected profit (solid), expected demand (bold), and price setting rule p^* (dashed).

Notice that, under the above parameterization, the solutions of the decision problem of a firm expecting q_t with certainty is almost identical to the solutions of the uncertain firm with expected shift parameter q_t: (5.9) and (5.10) are identical to (5.14) and (5.15) if both parameters $\gamma(\alpha, \eta)$ and $c(\alpha, \eta)$ are set equal to one. Moreover, the updating equations (5.8) and (5.12) are identical. The following analysis will be restricted to the (more general) Bayesian case, while the case of adaptive expectations can be obtained by setting the parameters γ and c equal to one.

5.3.2 A reduced dynamical system

In order to simplify the analysis, neglect expectations of households by assuming a Cobb-Douglas utility function and by ignoring profit income expectations ($\pi_t^e = 0$) which implies that labor supply is constant: $L^s(w_t, p_{t+1}^e, \pi_t^e) \equiv \bar{L} < L_{\max}$. Also abstract from inventory holdings by choosing $\delta = 1$.

The wage adjustment rate is specified by

$$\mathcal{W}(\bar{L}, L^d) = \mu_w \frac{L^d - \bar{L}}{\max(L^d, \bar{L})} \quad , \ \mu_w \in [0,1] \ .$$

It is immediate to see that all demands and supplies, namely $L^d_{t,i}$, $x^s_{t,i}$, and $y_{t,i}$, $i = 1, \dots, n$, are homogeneous of degree zero in M_t, w_t and $q_{t,i}^{1/\eta_i}$, $i = 1, \dots, n$. Thus, with a normalization with respect to labor

$$m_t := \frac{M_t}{w_t} \quad \text{and} \quad \zeta_{t,i} := \frac{q_{t,i}^{1/\eta_i}}{w_t}$$

it is clear that $L^d_{t,i}$, $x^s_{t,i}$, and $y_{t,i}$ may be written as functions depending only on m_t and $(\zeta_{t,i})_{i=1,\dots,n}$. From (5.14) it follows that $p_{t,i}/w_t$ and thereby also the sets I^l_t depend only on m_t and $(\zeta_{t,i})_{i=1,\dots,n}$.

Define $l_{t,i} := l$ if $i \in I^l_t$ to be number of the market (ordered by low prices) on which firm i trades, and let $n_{t,i} := \#I_t^{l_{t,i}}$ be the number of firms on this market. Firm i gets then a share of the government consumption equal to $g_t^{l_{t,i}}/n_{t,i}$. This implies that the $n+2$–dimensional dynamical system (5.4), (5.5) and (5.12) can be reduced to the $n+1$–dimensional system in $(m, \zeta) \in \mathbb{R}^{n+1}_{++}$: [4]

$$m_{t+1} = \Big(m_t + \sum_{i=1}^{n} \zeta_{t,i} \Big(\frac{\gamma(\alpha_i, \eta_i)}{x^s_{t,i}}\Big)^{1/\eta_i} \Big(\frac{g_t^{l_{t,i}}}{n_{t,i}} - \tau x_{t,i}\Big)\Big) \Big(1 + \mathcal{W}(\bar{L}, L^d_t)\Big)^{-1} \tag{5.16}$$

$$\zeta_{t+1,i} = \zeta_{t,i} \Big(b_i + (1 - b_i)\gamma(\alpha_i, \eta_i)\frac{y_{t,i}}{x^s_{t,i}})\Big)^{1/\eta_i} \Big(1 + \mathcal{W}(\bar{L}, L^d_t)\Big)^{-1} \ . \tag{5.17}$$

5.4 Analysis of the dynamics

5.4.1 The monopolistic case

In case of $n = 1$, (5.16) and (5.17) define a two-dimensional dynamical system. First, stationary behavior will be classified. As in case of com-

[4] That is, the trajectories of the reduced dynamical system describe the same allocations as the trajectories of the full system with corresponding initial values.

petitive price adjustment (cf. Böhm, Lohmann & Lorenz (1994)), disequilibrium steady states must be of Keynesian Unemployment or Repressed Inflation, depending on the budget deficit at full employment. Only in case of a balanced budget, there is a market clearing steady state. However, one has to assume that the monopolist produces its expected demand since otherwise different goods market disequilibria may occur.

Theorem 1: Let $(m, \zeta) \gg 0$ be a steady state. If the monopolist has adaptive expectations or Bayesian expectations with $\eta = \hat{\eta}(\alpha)$, and if $\mu_w > 0$, then $(m, \zeta) \in \mathbf{K}$ iff $g < \tau F(\bar{L})$, $(m, \zeta) \in \mathbf{M}$ iff $g = \tau F(\bar{L})$, and $(m, \zeta) \in \mathbf{I}$ iff $g > \tau F(\bar{L})$.

Proof:
Let (x, L) be the allocation on the goods and labor market resulting from (m, ζ), and let L^d, y, x^s be the corresponding labor demand, goods demand and goods supply.
Assume first $g > \tau F(\bar{L})$. This implies $g - \tau x \geq g - \tau F(\bar{L}) > 0$. Stationarity of m implies then that $\mathcal{W}(\bar{L}, L^d) > 0$ and therefore $L^d > \bar{L}$. Stationarity of ζ implies that $b + (1 - b)\gamma(\alpha, \eta)\frac{y}{x^s} > 1$ which under $\gamma(\alpha, \eta) = 1$ implies that $y > x^s = x$. Thus, $(m, \zeta) \in \mathbf{I}$.

Assume now $g = \tau F(\bar{L})$. Then $g - \tau x \geq g - \tau F(\bar{L}) = 0$. If this inequality was strict, stationarity of m would imply as above $L^d > \bar{L}$ and $y > x = x^s = F(\bar{L})$, a contradiction. Therefore, $g = \tau x$, from which by stationarity of m and ζ it follows that $\bar{L} = L^d$ and $y = x^s$, and therefore $(m, \zeta) \in \mathbf{M}$.

Lastly, assume $g < \tau F(\bar{L})$. If $g - \tau x \geq 0$, it follows as above by stationarity of m and ζ that $L^d \geq \bar{L}$ and $y \geq x = x^s = F(\bar{L})$, from which the contradiction $g - \tau F(L_{\max}) \geq 0$ follows. This implies $g - \tau x < 0$, from which by stationarity of m and ζ it follows that $y < x^s$ and $L^d < \bar{L}$ or $(m, \zeta) \in \mathbf{K}$. □

Next, global stability of a steady state will be investigated. Assume first that wage rates are stationary ($\mu_w = 0$) and either that there is no government activity or that government chooses a policy which does not effect aggregate demand, i.e. assume that the aggregate demand function $M/p + g$ is time-independent. [5] In this case, it suffices to analyze the one-dimensional dynamical system described by equation (5.12). Simulations show that the stability result of the following Theorem remains also true under slow wage adjustment (small $\mu_w > 0$) and fixed government parameters.

Theorem 2: Let $B \leq 1$ (non-increasing returns), $\mu_w = 0$ and $M/p + g$ be the time-independent aggregate demand. If $\bar{q}$ is a steady state under (5.12), then $\bar{q}$ is globally stable.

Proof: (5.15) and (5.3) imply that goods supply may be written as

$$x^s(q) := \min\left(F(\bar{L}), c_1 q^{\frac{B}{\eta(1-B)+B}}\right)$$

with some constant $c_1 > 0$. (5.14) implies then that there are $c_2, c_3 > 0$ such that

$$p^*(x^s(q), q) = \max\left(c_2 q^{1/\eta}, c_3 q^{\frac{1-B}{(1-B)\eta+B}}\right) \quad .$$

Therefore, there exists $\hat{q}$ such that p^* is differentiable in $q \neq \hat{q}$ and there are constants $\lambda_1, \lambda_2 \leq 1/\eta$ such that

$$\frac{d}{dq} p^*(x^s(q), q) = \lambda_i \frac{p^*(x^s(q), q)}{q} \quad , \forall\, q \neq \hat{q}.$$

Differentiation of (5.12) gives

$$\frac{dq_{t+1}}{dq_t}(q) = b + (1-b)\left(Mp^{\eta-2}(\eta-1) + g\eta p^{\eta-1}\right)\frac{p}{q}\lambda_i > 0 \ , \ q \neq \hat{q},$$

with $p = p^*(x^s(q), q)$. In a steady state $\bar{q}$ and $\bar{p} = p^*(x^s(\bar{q}), \bar{q})$ it is $\bar{q} = M\bar{p}^{\eta-1} + g\bar{p}^{\eta}$ and it follows

$$\frac{dq_{t+1}}{dq_t}(\bar{q}) < b + \lambda_i \eta (1-b) \leq 1$$

[5] This can be achieved by a balanced–budget tax policy $\tau_t = \frac{g_t}{x_t}$, where g_t are real government expenditures and x_t is the output (which equals the real income).

if $\bar{q} \neq \hat{q}$. Moreover, in $\hat{q}$ the iteration is continuous and the left and right derivative is strictly positive. In the special case, $\hat{q} = \bar{q}$ both derivatives are strictly less than one. Thus, the steady state is unique and globally stable. □

This stability result remains not true if the technology has increasing returns to scale. Such a result has been shown by Maußner (1996) in a different model with monopolistic competitors and adaptive expectations. The one-dimensional difference equation $q_t \mapsto q_{t+1}$ may be no longer monotonically increasing and may have two critical points with very steep slope between them. One finds that a period-doubling cascade occurs for increasing B up to $\eta/(\eta-1)$. In the full two–dimensional model, period-doubling bifurcations may also occur if μ_w is increased, since high adjustment speeds in price adjustment processes may lead to overshooting–effects, see e.g. Weddepohl (1996). Under decreasing returns to scale, μ_w would have to be very large (close to 1) to get instability. However, choosing such high wage adjustment speeds is uncommon, since wages are typically assumed to change slowly. Nevertheless, under strong enough increasing returns even for $\mu_w = 0$ instability may occur.

In Figure 5.3 a (possibly strange) attractor (i.e. the last 4500 of 5000 iterations) is plotted in the case of a Bayesian monopolist with increasing returns ($B = 1.8$, $\eta = 2$), no government activity ($g = \tau = 0$) and moderate wage adjustment ($\mu_w = 0.2$). The picture illustrates the long–run behavior of the dynamical system (5.16) and (5.17), and it indicates the frequency of the occurrence of different rationing situations. One observes that all types of disequilibria occur with the exception of Repressed Inflation. Since $\alpha = 2$ and $\eta = \hat{\eta}(2) = 2$, the unique steady state is at the market clearing situation. One should further notice that this instability is not robust under government activity. One finds that balanced budget policies with low values of government demand may stabilize market clearing steady states. This can be explained by a shift of aggregate demand to the right which has a positive effect on $\frac{dq_{t+1}}{dq_t}$ (compare the proof of Theorem 2).

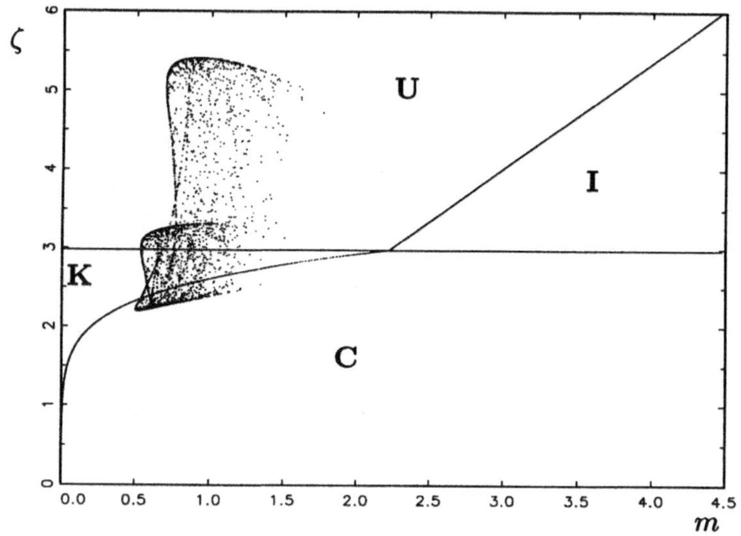

Figure 5.3: An attractor under increasing returns.

5.4.2 The oligopolistic case

Consider now the case $n \geq 2$, but assume that all firms have decreasing returns ($B_i < 1$) in order to exclude instabilities of the type analyzed in the previous section. The most important difference between the monopolistic and the oligopolistic case is the influence of decisions of firms on the outcomes and expectations of the other firms. The following exposition will concentrate on this interaction and, for the moment, abstract from wage adjustment ($\mu_w = 0$) and government activity ($g = \tau = 0$), so that the dynamics is described by (5.12) only.

Since there is a homogeneous output, demand functions are discontinuous which implies that the dynamics has discontinuities whenever some firms set equal prices. Furthermore, in case of completely symmetric firms it will turn out that the single steady state is a point of discontinuity which implies that it must be unstable (this is a trivial consequence from the definition of continuity). The next Theorem will show that no steady state exists with continuous dynamics in its neighborhood and that there always exists an unstable steady state with all firms setting the same price. First, the following notation is needed: Given a vector of expected shift parameters $\hat{q} = (q_i)_{i=1,\dots,n}$, write $x_i^s(\hat{q})$ for the supply of each firm (after possible labor market rationing), $p_i(\hat{q})$ for its price, and $y_i(\hat{q})$ for its demand.

Theorem 3: Assume $\mu_w = g = \tau = 0$. Let $\alpha_i = \alpha$, $\eta_i = \eta$, $A_i = A$ and $B_i = B < 1$ for $i = 1, \dots, n$. Then there exists no steady state $\hat{q}$ with $p_i(\hat{q}) \neq p_j(\hat{q})$ for all $i \neq j$. Furthermore, there exists a unique unstable steady state $\hat{q}$ with $p_i(\hat{q}) = p_j(\hat{q})$ for all i, j.

Proof: Suppose that there exists a steady state $\hat{q}$ with $p_1(\hat{q}) < \dots < p_n(\hat{q})$. Observe first that because of complete symmetry between firms, the supply of firms is ordered in the same way as demand expectations, i.e. $x_i^s(\hat{q}) < x_j^s(\hat{q})$ iff $q_i < q_j$.[6] Secondly, observe that prices are ordered in the same way, i.e. $p_i(\hat{q}) < p_j(\hat{q})$ iff $q_i < q_j$ (this follows from (5.15), (5.14), and proportional rationing). Therefore, the assumed ordering of prices implies that

$$x_1^s(\hat{q}) < x_2^s(\hat{q}) \quad . \tag{5.18}$$

Furthermore, $p_1(\hat{q}) < p_2(\hat{q})$ implies

$$y_2(\hat{q}) = \frac{M - p_1(\hat{q})x_1^s(\hat{q})}{p_2(\hat{q})} < \frac{M}{p_1(\hat{q})} = y_1(\hat{q}) \quad . \tag{5.19}$$

On the other hand, stationarity of q_1 and q_2 implies

$$\frac{x_1^s(\hat{q})}{y_1(\hat{q})} = \gamma(\alpha, \eta) = \frac{x_2^s(\hat{q})}{y_2(\hat{q})} \quad ,$$

which is a contradiction to (5.18) and (5.19).

[6] This follows from (5.15). If there is a binding labor demand constraint, it uses proportional rationing.

In order to prove the existence of a completely symmetric steady state, write $\hat{q} = (q)_{i=1,\dots,n}$. Stationarity of q is equivalent to $x_i^s(\hat{q}) = \gamma(\alpha,\eta)y_i(\hat{q})$. Since all q_i are identical, one has $y_i(\hat{q}) = M/(np(\hat{q}))$. Therefore, stationarity of q is equivalent to

$$M\gamma(\alpha,\eta)\left(\frac{x^s(q)}{\gamma(\alpha,\eta)q}\right)^{1/\eta} = nx^s(q) \tag{5.20}$$

where the symmetric supply may be written (with some $c > 0$)

$$x^s(q) = \min\left(F(\bar{L}/n), cq^{\frac{B}{(1-B)\eta+B}}\right) \quad .$$

The left hand side in (5.20) is monotonically decreasing in q, tends to infinity for $q \to 0$ and tends to zero for $q \to \infty$, while the right hand side is strictly positive for $q > 0$ and non-decreasing. Therefore, (5.20) has a unique solution q. Instability of the unique symmetric steady state follows from the observation that a small deviation of q_i below q forces firm i to deviate with its price below and to get the whole market demand $M/p_i \gg M/(np)$ which implies that q_i in the next period is strictly larger than the steady state value. □

This Theorem implies that one cannot find a locally stable steady state in a symmetric oligopoly. In the general case, it is furthermore clear that a locally stable steady state requires that all firms set different prices. This would imply that the firm with the lowest price supplies less than its demand (otherwise residual demand of higher price firms cannot be positive which is necessary for stationarity).

Proposition 2: Assume $\mu_w = g = \tau = 0$. Let $\hat{q}$ be a locally stable steady state in the model with Bayesian oligopolists. Then, for some i, $\eta_i > \hat{\eta}(\alpha_i)$.

Proof: If $\hat{q}$ is a locally stable steady state, firms must charge different prices. Because residual demand at higher price firms must be positive, there exists some firm i supplying less than demand, i.e. $x_i^s(\hat{q}) < y_i(\hat{q})$.

Stationarity of q_i and (5.12) yield $y_i(\hat{q}) = x_i^s(\hat{q})/\gamma_i = q_i p_i(\hat{q})^{-\eta_i}$. Thus, firm i supplies also less than expected demand, and Proposition 1 yields $\eta_i > \hat{\eta}(\alpha_i)$. □

In order to get an impression of the global dynamics, restrict for the moment to the case $n = 2$. In the symmetric case, it turns out that mostly symmetric cycles of period two occur. To see this, note that $q_{t+1,i} < q_{t,i}$ if and only if $q_{t,i} p_i(\hat{q}_t)^{-\eta_i} > y_i(\hat{q}_t)$. Expected demand turns out to be continuous and monotonically increasing in q_i, while demand is the discontinuous function

$$y_i(\hat{q}) = \begin{cases} \dfrac{M}{p_i(\hat{q})} & , \quad p_i(\hat{q}) < p_j(\hat{q}) \, , \\ \left(\dfrac{M}{p_i(\hat{q})}\right) x_i^s(\hat{q})/(x_i^s(\hat{q}) + x_j^s(\hat{q})) & , \quad p_i(\hat{q}) = p_j(\hat{q}) \, , \\ \max\left(0, \dfrac{M - p_j(\hat{q}) x_j^s(\hat{q})}{p_i(\hat{q})}\right) & , \quad p_i(\hat{q}) > p_j(\hat{q}) \, . \end{cases}$$

Figure 5.4 illustrates qualitatively the dynamics in the symmetric duopolistic case with no binding labor market constraint (i.e. $\bar{L}$ and w sufficiently large). In this case x_i^s and p_i are independent of q_j for $j \neq i$, such that for $q_i < q_j$ the stationarity condition of q_i depends only on q_i, since firm i sets the lowest price and gets the whole demand. For $q_i > q_j$ stationary values of q_i increase if q_j decreases, since in this case firm j supplies less and leaves firm i a higher residual demand. Along the diagonal $q_1 = q_2$ firms set equal prices and the dynamics is discontinuous. Consequently, there are jumps in both loci of stationarity. In case of $\eta = \hat{\eta}(\alpha)$ or in case of adaptive expectations, the steady state is at the intersection of D_1, D_2, E_1 and E_2. Otherwise the steady state could be above or below on the diagonal. The arrows indicating the change of q_1 and q_2 show that there will most likely occur a symmetric two-cyclical dynamics between D_1 and D_2. Simulations show that in fact symmetric cycles of period 2 with alternating prices occur.

However, the situation changes dramatically if firms are asymmetric. In this case, the locus of discontinuity is no longer a diagonal and, more

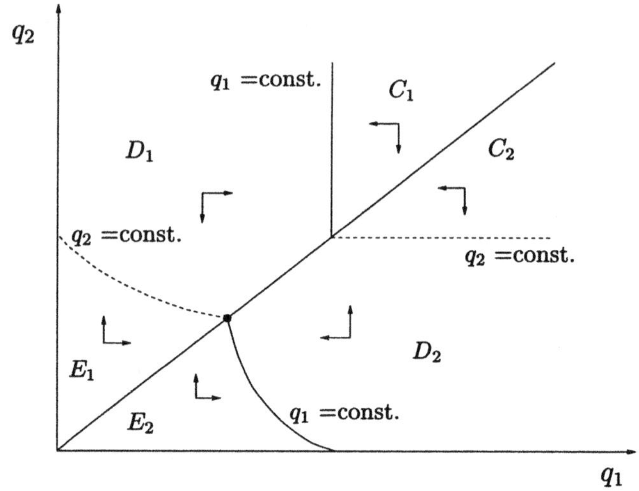

Figure 5.4: Qualitative dynamics of a symmetric duopoly.

importantly, the loci of stationarity of q_i are displaced against each other, leaving therefore a greater variety of cycles between different regions. For instance, a small deviation in the subjective price elasticity of one firm suffices to generate cycles of different periods. As an example, Figure 5.5 shows a cycle of period 9 in case of Bayesian updating with $\alpha = 2$, $\eta_1 = 2.1$, $\eta_2 = 2$ and $B_1 = B_2 = 0.5$. Picture (a) shows that profits of both firms fluctuate, but that no firm makes losses. Firm 1, which is subjectively more competitive, has higher profits on average. The fluctuating prices in picture (b) show that firms alternately undercut each other, implying that the demand of both firms in pictures (c) and (d) alternately falls above and below supply with demand having a greater volatility than supply.

In order to illustrate the variety of cycles for different subjective price elasticities, Figure 5.6 shows for different combination of $(\eta_1, \eta_2) \in [1, 3]$ colors which correspond to cycles of certain periods.[7] The corresponding

[7]A colored copy of this figure can be obtained from the author or from http://w3.ihs.ac.at/~kaas/diss.html . The pictures were computed and drawn using MACRODYN, a package for the simulation of discrete–time dynamical systems, see

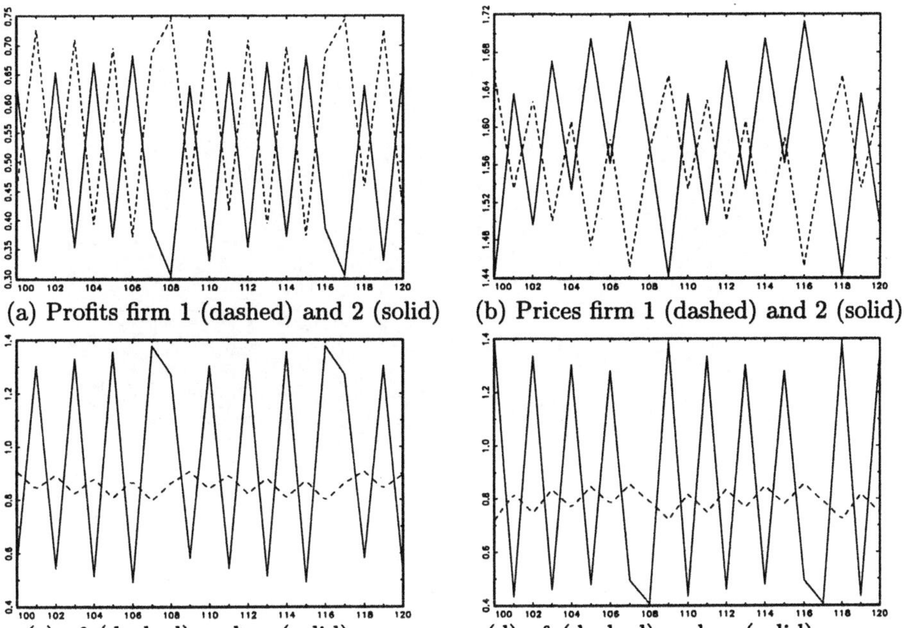

(a) Profits firm 1 (dashed) and 2 (solid)
(b) Prices firm 1 (dashed) and 2 (solid)
(c) x_1^s (dashed) and y_1 (solid)
(d) x_2^s (dashed) and y_2 (solid)

Figure 5.5: Period-9-cycle in a duopoly with $\eta_1 = 2.1$ and $\eta_2 = 2$.

color code is indicated below. Picture (a) shows the case of Bayesian learning with $\alpha = 2$ ($b = 0.5$) and $B = 0.5$: Along the diagonal there are cycles of period 2. For strong asymmetries and $\eta_i > \hat{\eta}(\alpha) = 2$ for some i, a steady state occurs where the lower price firm permanently supplies less than demand. Picture (b) shows the case of adaptive expectations for the same parameter set. As discussed above, stable steady states cannot be observed since it does not happen that some firm supplies permanently less than demand. Instead the occurrence of cycles of period 2 is much more likely than in the case of Bayesian updating.

Both pictures show that stable cycles of a very high order may occur. In these cases it seems plausible that firms do not recognize these cycles and could misunderstand them as stochastic fluctuations. This seems the more likely the higher the order of the cycle is. Of course, it might still

Schenk-Hoppé et al. (1998).

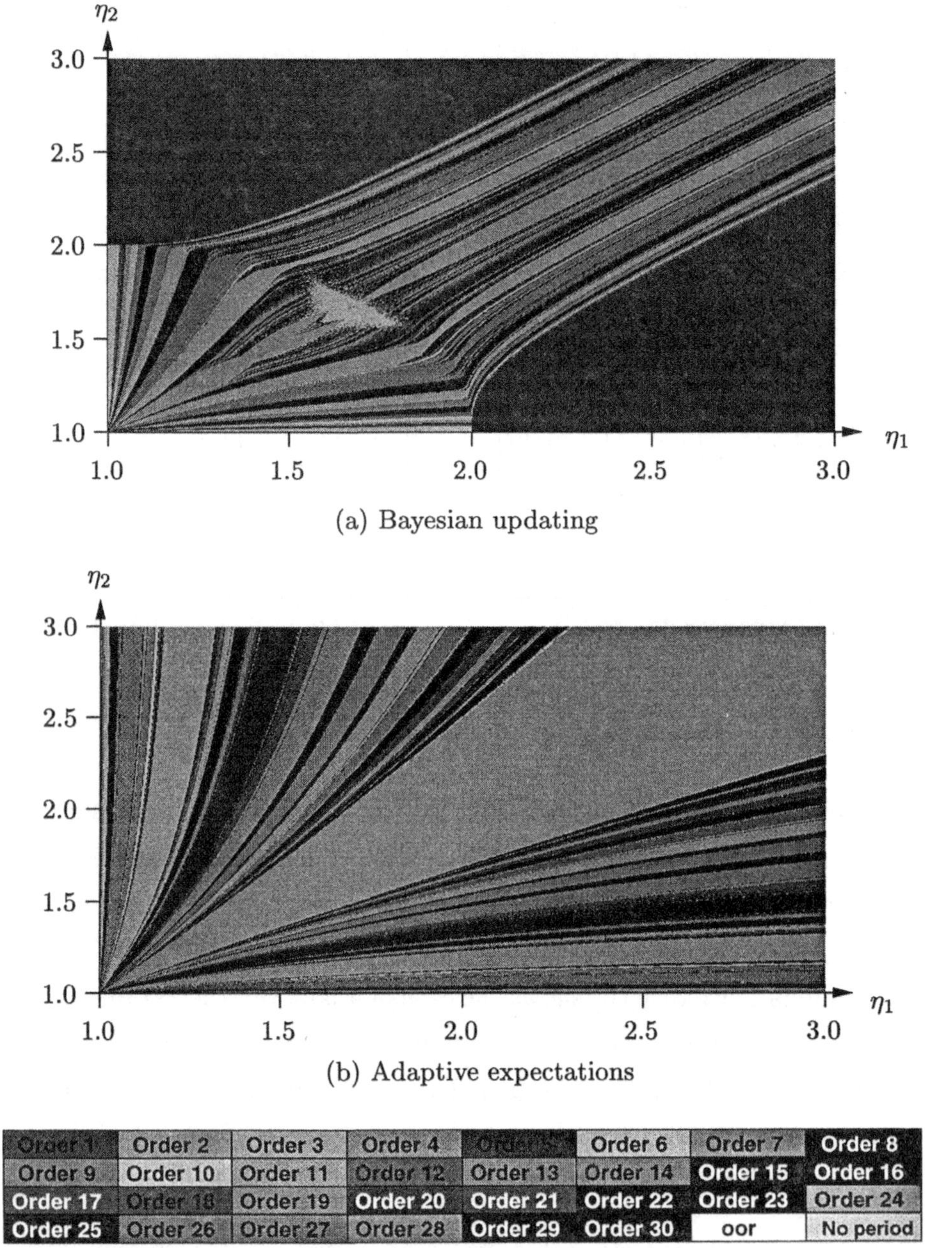

(a) Bayesian updating

(b) Adaptive expectations

Figure 5.6: Two-parameter bifurcations with color code.

be that some lower order autocorrelation coefficients are different from zero. This seems to be the case in Figure 5.5. Thus, the simple model of the firms is probably not consistent in the sense of Hommes & Sorger (1998).

One may ask, however, what happens if the firms use another learning scheme which is capable of recognizing cycles up to a certain order. For instance, firms may have a memory of length T and test it for cycles up to the order $\tau < T$. If a cycle of a certain order is recognized by price setters, it would not persist anymore, since price setters would choose their supply equal to the demand level, and leave no positive residual demand for higher price firms. This has been checked numerically in this model with 2 symmetric firms who are capable of recognizing cycles up to order 20. If, at the beginning, both firms have "naive expectations" (1–cyclical expectations), a cycle of period 2 occurs. After some time, agents learn the cycle, and a cycle of period 4 emerges. When this cycle is recognized, cycles of period 8, 16, and 32 emerge successively. The cycle of period 32 persists, since agents believe still in the cycle of period 16 and are not capable of recognizing the cycle of order 32. Similar observations of such a "period doubling" over time can be made in asymmetric cases.

One might expect that for $n > 2$ symmetric firms cycles of period n occur. However, this turns out not to be the case. Instead one finds that there are usually m groups of size l_k, $k = 1, \ldots, m$, $(\sum_{k=1}^{m} l_k = n)$ where prices, demands and supplies in each group k cycle with period l_k. This implies that the period of the common cycle is equal to the lowest common multiple of the l_k, say p_n. Table 5.1 shows the cycles occurring at different n for $\alpha = 2$, $\eta = 2$, $B = 0.5$, $\bar{L} = 10$, $w = 1$, $\mu_w = 0$, $M = 2$. All the cycles turned out to be symmetric which means that each cycle is generated by iteration of one permutation of firms. The symmetry group of the dynamical system is the permutation group of the n firms Π_n which is much larger than the symmetry group of one particular cycle which has order p_n. Thus, there is typically a very large number $\#\Pi_n/p_n$ of coexisting cycles which differ only with respect to permutations of firms.

n	Cycles	Period	L	n	Cycles	Period	L
2	2	2	0.338	20	3;7;10	210	0.92
3	3	3	0.385	30	4;6;9;11	396	1.131
4	2;2	2	0.441	40	3;4;8;25	600	1.313
5	1;4	4	0.477	50	4;5;6;8;13;14	10920	1.474
6	6	6	0.52	60	4;10;19;27	10260	1.621
7	2;5	10	0.552	70	28;42	84	1,756
8	4;4	4	0.591	80	19;61	1159	1.888
9	9	9	0.625	90	10;19;61	11590	2.014
10	2;8	8	0.655	100	5;19;76	380	2.132

Table 5.1: Cycles and employment for increasing number of firms.

Since all cycles are symmetric, aggregate employment along a cycle is constant. The corresponding employment levels are also indicated in Table 5.1. One finds that employment increases when the number of firms increases. It is interesting that employment increases above the steady state employment level, and even above the employment level with perfect or Cournot competition on the commodity market. The Cournot employment level for the above parameter values can be computed as $1 - 1/n < 1$. The steady state employment level turns out to be $c(\alpha, \eta)(1 - 1/\eta) < 1$. The explanation of this phenomenon is simple: Consumers, visiting firms with different prices and being rationed at low prices, express in the aggregate a higher demand than they are able to pay. Thus, average and expected demand of a single firm are usually higher than its corresponding share of aggregate demand, which drives firms to employ more labor than in a one-price economy. As a consequence, labor market clearing wages are higher in such a cycling economy than in a one-price economy.

One may ask, however, whether there really exists a competitive wage adjustment mechanism leading to permanent full employment. The answer turns out to be dependent on symmetry assumptions. With symmetric firms, symmetric cycles occur so that aggregate employment at a given wage rate is constant. In this case, the competitive wage adjustment mechanism (5.4) results in full employment. Figure 5.7 shows such a situation for $n = 10$, $\bar{L} = 1$ and $\mu_w = 0.1$. Even though demand and supply of a single firm fluctuate (in picture (a) firm 1 is in a cycle of period 8), employment converges to the full employment level in picture (b).

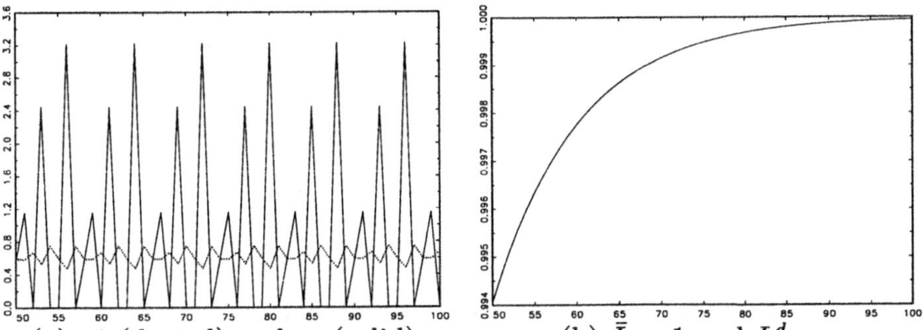

(a) x_1^s (dotted) and y_1 (solid) (b) $\bar{L} = 1$ and L^d

Figure 5.7: Demand and supply time series for 10 symmetric firms.

The situation is different in the asymmetric cases. If the 10 firms have subjective demand elasticities uniformly distributed between 2 and 2.1, a cycle of period 60 occurs, with fluctuating employment levels at a constant wage rate. The wage adjustment mechanism is not able to achieve full employment, and situations of excess labor demand and supply alternate (Figure 5.8). This result reveals that a competitive labor market and an imperfectly competitive goods market result in temporary unemployment, provided that firms are sufficiently asymmetric.

Furthermore, it turned out that firms with the highest prices typically get zero demand. In the definition of the disequilibrium regimes in Section 5.2, this means that there is permanent excess supply on the goods

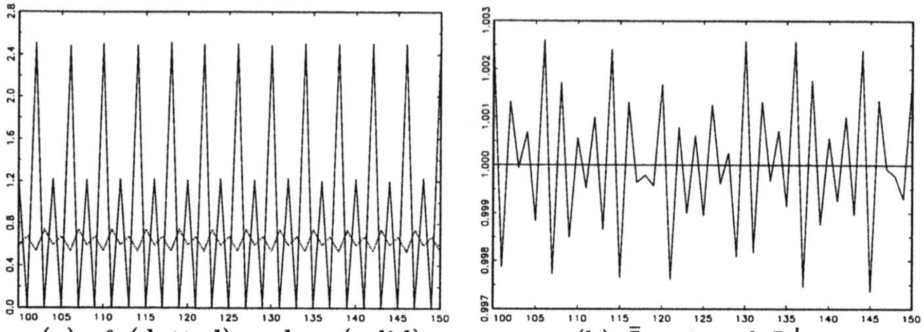

(a) x_1^s (dotted) and y_1 (solid) (b) $\bar{L} = 1$ and L^d

Figure 5.8: Demand and supply time series for 10 asymmetric firms.

market. Thus, the fluctuating dynamics for asymmetric firms alternates between **K** and **U**. Under government activity other rationing situations may certainly be possible.

5.5 Summary and discussion

In this chapter a simple dynamic macroeconomic model with imperfect competition has been analyzed. Firms set prices on the goods market at subjective demand expectations which are updated by gaining new demand information. On the labor market there is sluggish but competitive wage adjustment. It turns out that in the case of a single monopolistic firm with convex technology the unique steady state is globally stable, but that cyclical and irregular dynamics may occur if the monopolist has strongly increasing returns. If there are many firms, even under decreasing returns cyclical dynamics can be observed which shows increasing variety under small asymmetries between firms. Such asymmetries are shown to be also the cause of the failure of the competitive wage adjustment mechanism.

These cycles under price/quantity competition of n suppliers of a homogeneous product may be interpreted as "Edgeworth cycles", with the difference that they occur in a really dynamic model and not as mixed strategy Nash equilibria in a one–shot game, which seems to be more in accordance with the ideas of Edgeworth, as Tirole (1989) emphasizes:

> "*Thus, mixed strategies call for price dynamics. Indeed, when Edgeworth introduced capacity constraints to escape the Bertrand paradox, he suggested the possibility of price cycles rather than the use of mixed strategies.*"

Clearly, these results rest on many simplifying assumptions, and one can think of many important extensions, which are briefly sketched in the following. Whatever extensions or modifications are introduced in the model, the dynamics must be expected to be unstable, as long as the demand functions have discontinuities.

From a macroeconomic point of view, the homogeneous good assumption seems to be a too strong simplification. It must be noted, however, that an extension to the case of many differentiated products is not straightforward, if only voluntary trade is allowed to take place and no fixed price equilibrium should be employed. One possibility might be to let consumers visit the different firms within each period in a given order as long as they do not want or cannot trade anymore.

Another simplification of the above model is that prices and wages are assumed to be fixed for the same length of period. The model provides no microeconomic explanation for prices and wages to be fixed for a certain time, but wages are typically regarded to be more rigid than prices (due to wage contracts, for instance). One could either circumvent the problem by introducing large enough menu costs for price changes within a period, or modify the model by allowing firms to change prices several times. In the latter case, however, one has to take into account the possibility of price revisions in the expectation structure of price setters as well as in the demand behavior of consumers.

Further important extensions concern the expectation structure of price setters. First, firms use here a very simple model of demand which could be replaced by more sophisticated models estimating more parameters of an unknown demand function and using more data than those one period before. Second, firms do not think strategically in the sense that their perception of demand includes expectations and past information about actions of competitors. Especially in the case of a homogeneous oligopoly with very few firms such an assumption seems implausible. An explicit consideration of other firms' actions would require a splitting of each firm's subjective demand function (i.e. its IOF) in an action expectation function and an outcome expectation function depending on all firms' actions, as it has been formulated in Chapter 2.

Chapter 6

Final remarks

In this thesis, I have examined different approaches to dynamic macroeconomic theory with imperfect competition. The general framework introduced in Chapter 2 allows a characterization of different equilibrium concepts as stationary states of a dynamic process. It has been argued that the objective demand approach to dynamic macroeconomics must be questioned for the following reasons:

First, as discussed in Chapters 2 and 3, the objective demand approach does not appear to be the appropriate concept of a short–run equilibrium, since the existence depends crucially on the specifications of the economy. Furthermore, even if objective equilibria exist, they are not more plausible limit outcomes of an adjustment process than other subjective equilibria. In some cases objective demand expectations may actually be more unrealistic than subjective ones.

Second, even if an objective equilibrium was preferred as the appropriate short–run equilibrium concept, it has been shown in Chapter 4 that this approach would not be able to characterize unique long–run (stationary) equilibria. That is, intertemporal equilibria, in which all agents have rational expectations and in which all imperfect competitors maximize against their objective demand curve, are completely indeterminate.

An alternative and strictly dynamic model has been presented in Chapter 5. All markets open sequentially, and price setting firms believe in subjective demand functions. They try to learn the parameters of these functions by making observations of past demands that they have been faced with. The major advantage of this approach is that the model can be formulated for arbitrary specifications of preferences, technologies and beliefs, since no equilibrium concept is used in the definition of the dynamic process. The only requirement one has to impose on the model specification is that the individual decision problems must have solutions, which is a fairly weak restriction. Although the basic ideas of this approach were already discussed more than twenty years ago, it is still in its infancy and should in the future be extended in many respects. Most important in my opinion are the extensions to differentiated products, to other time structures of price and wage adjustment, and to other demand expectations of price setters, as has been sketched at the end of Chapter 5.

One might still raise the objection that the subjective demand approach leaves too much arbitrariness in the model, since stationary equilibria as well as the dynamic evolution depend critically on the beliefs of agents and on the way how they update their beliefs. While it is probably true that it is difficult to obtain general results, the criticism concerning the indeterminacy of subjective demand expectations can best be replied by the following remark of Kaldor (1934):

> "*The 'real demand curve' confronting the individual producer might be (...) indeterminate; (...) The 'imagined demand curve', on the other hand, becomes determinate as soon as it exists in the producer's imagination – and since something always must exist there the question of indeterminateness simply does not arise in this case. If on the other hand, by assuming perfect knowledge we make the two coincide, not only do we make the analysis unnecessarily unrealistic, but*

we introduce complications (by rendering the 'imagined demand curve' indeterminate) which can be avoided."

List of Figures

List of Abbreviations

CES	Constant elasticity of substitution
IE	Intertemporal equilibrium
IOF	Individual outcome function
OLG	Overlapping generations
OLS	Ordinary least squares
PFE	Perfect foresight equilibrium/equilibria
RBC	Real business cycle
REE	Rational expectations equilibrium/equilibria
TSCE	Temporary symmetric Cournot equilibrium/equilibria

Bibliography

ARROW, K. (1959): "Towards a Theory of Price Adjustment", in *The Allocation of Economic Resources*, ed. by A. Abramovitz. Stanford University Press, Stanford.

ARROW, K. & F. HAHN (1971): *General Competitive Analysis*. Holden–Day, San Francisco.

BALL, L. & D. ROMER (1989): "The Equilibrium and Optimal Timing of Price Changes", *Review of Economic Studies*, 56, 179–198.

BARRO, R. & H. GROSSMAN (1971): "A General Disequilibrium Model of Income and Employment", *American Economic Review*, 61, 82–93.

BENASSY, J.-P. (1975): "Neokeynesian Disequilibrium Theory in a Monetary Economy", *Review of Economic Studies*, 42, 503–524.

——— (1976): "The Disequilibrium Approach to Monopolistic Price Setting and General Monopolistic Equilibrium", *Review of Economic Studies*, 43, 69–81.

——— (1977): "A Neokeynesian Model of Price and Quantity Determination in Disequilibrium", in *Equilibrium and Disequilibrium in Economic Theory*, ed. by G. Schwödiauer. Reidel, Boston.

——— (1982): *The Economics of Market Disequilibrium*. Academic Press, New York.

——— (1988): "The Objective Demand Curve in General Equilibrium with Price Makers", *The Economic Journal*, 98, 37–49.

——— (1991): "Monopolistic Competition", in *Handbook of Mathematical Economics*, ed. by W. Hildenbrand & H. Sonnenschein, vol. IV, chap. 37. North–Holland, Amsterdam.

BERNANDO, J. & A. SMITH (1993): *Bayesian Theory.* Wiley, New York.

BLANCHARD, O. & N. KIYOTAKI (1987): "Monopolistic Competition and the Effects of Aggregate Demand", *American Economic Review*, 77, 647–666.

BÖHM, V. (1989): *Disequilibrium and Macroeconomics.* Basil Blackwell, Oxford.

——— (1994): "The Foundation of the Theory of Monopolistic Competition Revisited", *Journal of Economic Theory*, 63, 208–218.

——— (1998): "Macroeconomic Dynamics with Sequential Trading", Mimeo, University of Bielefeld.

BÖHM, V., M. LOHMANN & H.-W. LORENZ (1994): "Dynamic Complexity in a Keynesian Macroeconomic Model", Discussion Paper 288, Department of Economics, University of Bielefeld.

BÖHM, V., E. MASKIN, H. POLEMARCHAKIS & A. POSTLEWAITE (1983): "Monopolistic Quantity Rationing", *Quarterly Journal of Economics*, 98, 189–197.

BÖHM, V. & J. NAEVE (1995): "A Keynesian Macroeconomic Model with a Price Setting Firm", Mimeo, Department of Economics, University of Bielefeld.

BÖHM, V. & J. WENZELBURGER (1995): "Expectations, Forecasting, and Perfect Foresight - A Dynamical Systems Approach -", Discussion Paper 307, Department of Economics, University of Bielefeld, forthcoming in *Macroeconomic Dynamics.*

BONANNO, G. (1990): "General Equilibrium Theory with Imperfect Competition", *Journal of Economic Surveys*, 4, 297–328.

BONANNO, G. & E. ZEEMANN (1985): "Limited Knowledge of Demand and Oligopoly", *Journal of Economic Theory*, 35, 276–283.

CHIAPPORI, P. & R. GUESNERIE (1991): "Sunspot Equilibria in Sequential Markets Models", in *Handbook of Mathematical Economics*, ed. by W. Hildenbrand & H. Sonnenschein, vol. IV, chap. 32. North–Holland, Amsterdam.

D'ASPREMONT, C., R. DOS SANTOS FERREIRA & L. GÉRARD-VARET (1989): "Unemployment in an Extended Cournot Oligopoly Model", *Oxford Economic Papers*, 41, 490–505.

——— (1991): "Imperfect Competition, Rational Expectations, and Unemployment", in *Equilibrium Theory and Applications: Proceedings of the Sixth International Symposium in Economic Theory and Econometrics*, ed. by W. Barnett, B. Cornet, C. d'Aspremont, J. Gabszewicz & A. Mas-Colell, pp. 353–381. Cambridge University Press, Cambridge.

DEGROOT, M. (1970): *Optimal Statistical Decisions*. McGraw–Hill, New York.

DEHEZ, P. (1985): "Monopolistic Equilibrium and Involuntary Unemployment", *Journal of Economic Theory*, 36, 160–165.

DIERKER, E. & B. GRODAL (1995): "Profit Maximization, Relative Prices, and the Maximization of Shareholders' Real Wealth", Working Paper 9503, University of Vienna.

——— (1996): "The Price Normalization Problem in Imperfect Competition and the Objective of the Firm", Working Paper 9616, University of Vienna.

DIERKER, H. & B. GRODAL (1986): "Non–Existence of Cournot–Walras equilibrium in a General Equilibrium Model with Two Oligopolists", in *Contributions to Mathematical Economics in Honor of Gerard Debreu*, ed. by W. Hildenbrand & A. Mas-Colell. North–Holland, Amsterdam.

DIXON, H. & N. RANKIN (1994): "Imperfect Competition and Macroeconomics: A Survey", *Oxford Economic Papers*, 46, 171–199.

——— (1995): "Introduction", in *The New Macroeconomics: Imperfect Markets and Policy Effectiveness*, ed. by H. Dixon & N. Rankin. Cambridge University Press, Cambridge.

DRÈZE, J. H. (1975): "Existence of an Exchange Equilibrium under Price Rigidities", *International Economic Review*, 16, 301–320.

DUBEY, P. (1982): "Price–Quantity Strategic Market Games", *Econometrica*, 50, 111–126.

EDGEWORTH, F. (1897): "La Teoria Pura del Monopolio", *Giornale degli Economisti*, 40, 13–31, in English: The Pure Theory of Monopoly, in *Papers Relating to Political Economy*, ed. by F. Edgeworth, Macmillan Press, London (1925).

EICHBERGER, J. (1992): "Nash Equilibrium and Learning", in *Recent Developments in Game Theory*, ed. by J. Creedy, J. Borland & J. Eichberger, chap. 4. Elgar, Aldershot, U.K.

FELDERER, B. & S. HOMBURG (1987): *Macroeconomics and New Macroeconomics.* Springer–Verlag, Berlin.

FRIEDMAN, M. (1953): *Essays in Positive Economics.* University of Chicago Press, Chicago.

GABSZEWICZ, J. J. & J. P. VIAL (1972): "Oligopoly 'á la Cournot' in General Equilibrium Analysis", *Journal of Economic Theory*, 4, 381–400.

GARY-BOBO, R. J. (1989): "Cournot–Walras and Locally Consistent Equilibria", *Journal of Economic Theory*, 49, 10–32.

GAYGISIZ, E. & P. MADDEN (1997): "Macroeconomic Coordination Failure Under Oligempory", Discussion Paper 9721, School of Economic Studies, University of Manchester.

GERMANO, F. (1995): "Bertrand–Edgeworth Equilibria in Finite Exchange Economies", Discussion Paper 95-93, UC San Diego.

GRANDMONT, J.-M. (1982): "Temporary General Equilibrium Theory", in *Handbook of Mathematical Economics*, ed. by K. Arrow & M. Intriligator, vol. II, chap. 19. North–Holland, Amsterdam.

——— (1983): *Money and Value.* Cambridge University Press, Cambridge.

——— (1988): "Introduction", in *Temporary Equilibrium - Selected Readings*, ed. by J.-M. Grandmont. Academic Press, Boston.

——— (1998): "Expectations Formation and Stability of Large Socioeconomic Systems", *Econometrica*, 66, 741–781.

GRANDMONT, J.-M. & G. LAROQUE (1976): "On Temporary Keynesian Equilibria", *Review of Economic Studies*, 43, 53–67.

GRODAL, B. (1996): "Profit Maximization and Imperfect Competition", in *Economics in a Changing World*, ed. by B. Allen, vol. 2. MacMillan, London.

HAHN, F. H. (1978): "On Non–Walrasian Equilibria", *Review of Economic Studies*, 45, 1–17.

——— (1987): "Information Dynamics and Equilibrium", *Scottish Journal of Political Economy*, 34, 321–334.

HART, O. D. (1982): "A Model of Imperfect Competition with Keynesian Features", *Quarterly Journal of Economics*, 97, 109–138.

——— (1985): "Imperfect Competition in General Equilibrium: an Overview of Recent Work", in *Frontiers of Economics*, ed. by K. J. Arrow & S. Honkapohja, pp. 100–149. Basil Blackwell, Oxford.

HICKS, J. (1946): *Value and Capital.* Oxford University Press, Oxford, 2 edn.

HOMMES, C. & G. SORGER (1998): "Consistent Expectations Equilibria", *Macroeconomic Dynamics*, 2, 287–321.

HURWICZ, L. (1979): "Outcome Functions Yielding Walrasian and Lindahl Allocations at Nash Equilibrium Points", *Review of Economic Studies*, 46, 217–225.

IWAI, K. (1974): "The Firm in Uncertain Markets and Its Price, Wage and Employment Adjustments", *Review of Economic Studies*, 41, 257–276.

KAAS, L. (1997): "Imperfectly Competitive Price Setting under Bayesian Learning in a Disequilibrium Model", Discussion Paper 342, Department of Economics, University of Bielefeld.

——— (1998): "Multiplicity of Cournot Equilibria and Involuntary Unemployment", *Journal of Economic Theory*, 80, 332–349.

KALDOR, N. (1934): "Mrs. Robinson's 'Economics of Imperfect Competition' ", *Economica*, 1, 335–341.

KIEFER, N. M. & Y. NYARKO (1989): "Optimal Control of an Unknown Linear Process with Learning", *International Economic Review*, 30, 571–586.

KIRMAN, A. (1992): "Whom or What Does the Representative Individual Represent?", *Journal of Economic Perspectives*, 6, 117–136.

——— (1995): "Learning in Oligopoly: Theory, Simulation, and Experimental Evidence", in *Learning and Rationality in Economics*, ed. by A. Kirman & M. Salmon, pp. 127–178. Basil Blackwell, Oxford.

KREPS, D. & J. SCHEINKMAN (1983): "Quantity Precommitment and Bertrand Competition Yield Cournot Outcomes", *Bell Journal of Economics*, 14, 326–338.

LINDAHL, E. (1939): *Studies in the Theory of Money and Capital*. Allen & Unwin, London.

LUCAS, R. (1972): "Expectations and the Neutrality of Money", *Journal of Economic Theory*, 4, 103–124.

MADDEN, P. & J. SILVESTRE (1991): "Imperfect Competition and Fixprice Equilibria when Goods are Gross Substitutes", *Scandinavian Journal of Economics*, 93, 479–494.

MALINVAUD, E. (1977): *The Theory of Unemployment Reconsidered.* Blackwell, Oxford.

MANKIW, N. (1985): "Small Menu Costs and Large Business Cycles: A Macroeconomic Model of Monopoly", *Quarterly Journal of Economics*, 100, 529–539.

——— (1990): "A Quick Refresher Course in Macroeconomics", *Journal of Economic Literature*, XXVIII, 1645–1660.

MANKIW, N. & D. ROMER (1991): "Introduction", in *New Keynesian Economics*, ed. by N. Mankiw & D. Romer, pp. 1–26. The MIT Press, Cambridge, MA.

MARCET, A. & T. SARGENT (1992): "The Convergence of Vector Autoregressions to Rational Expectations Equilibria", in *Macroeconomics - A Survey of Research Strategies*, ed. by A. Vercelli & N. Dimitri. Oxford University Press, Oxford.

MARIMON, R. (1997): "Learning from Learning in Economics", in *Advances in Economics and Econometrics: Theory and Applications*, ed. by D. Kreps & K. Wallis. Cambridge University Press, Cambridge.

MARSCHAK, T. & R. SELTEN (1974): *General Equilibrium with Price–making Firms*, vol. 91 of *Lecture Notes in Economics and Mathematical Systems.* Springer–Verlag, Berlin.

MAS-COLELL, A., M. D. WHINSTON & J. R. GREEN (1995): *Microeconomic Theory.* Oxford University Press, Oxford.

MAUSSNER, A. (1996): "Monopolistic Price Setting, Subjective Demand, and the Business Cycle", in *Dynamic Disequilibrium Modeling,* ed. by W. Barnett, G. Gandolfo & C. Hillinger. Cambridge University Press, Cambridge.

MIRMAN, L. J., L. SAMUELSON & A. URBANO (1993): "Monopoly Experimentation", *International Economic Review*, 34, 549–563.

NEGISHI, T. (1961): "Monopolistic Competition and General Equilibrium", *Review of Economic Studies*, 28, 196–201.

——— (1972): *General Equilibrium Theory and International Trade.* North–Holland, Amsterdam.

——— (1979): *Microeconomic Foundations of Keynesian Macroeconomics.* North–Holland, Amsterdam.

NYARKO, Y. (1991): "Learning in Mis–specified Models and the Possibility of Cycles", *Journal of Economic Theory*, 55, 416–427.

PATINKIN, D. (1956): *Money, Interest and Prices.* Harper and Row, New York.

RAMPA, G. (1989): "Conjectures, Learning, and Equilibria in Monopolistic Competition", *Journal of Economics*, 49, 139–163.

RANKIN, N. (1992): "Imperfect Competition, Expectations and the Multiple Effects of Monetary Growth", *The Economic Journal*, 102, 743–753.

——— (1997): "Imperfect Competition and the Modelling of Expectations in Macroeconomics", prepared for A. Boitani and A. Salanti (eds.), Information and Coordination: A New Keynesian Perspective, Kluwer Publishers, Dordrecht.

ROMER, P. (1990): "Endogenous Technological Change", *Journal of Political Economy*, 98, 71–102.

ROTEMBERG, J. & M. WOODFORD (1992): "Oligopolistic Pricing and the Effects of Aggregate Demand on Economic Activity", *Journal of Political Economy*, 100, 1153–1207.

SCHENK-HOPPÉ, K. R. ET AL. (1998): "MACRODYN – A User's Guide", Mimeo, Department of Economics, University of Bielefeld.

SCHÖNHOFER, M. (1997): *Konsistentes adaptives Lernen in Finanzmarkt- und Makromodellen*, vol. 2151 of *Europäische Hochschulschriften*. Peter Lang, Frankfurt am Main.

SCHULTZ, C. (1992): "The Impossibility of Involuntary Unemployment in an Overlapping Generations Model with Rational Expectations", *Journal of Economic Theory*, 58, 61–76.

SHAPLEY, L. & M. SHUBIK (1977): "Trade Using a Commodity as a Means of Payment", *Journal of Political Economy*, 85, 937–968.

SILVESTRE, J. (1977): "A Model of General Equilibrium with Monopolistic Behavior", *Journal of Economic Theory*, 16, 425–442.

——— (1990): "There May be Unemployment When the Labour Market is Competitive and the Output Market is Not", *Economic Journal*, 100, 899–913.

——— (1993): "The Market–Power Foundations of Macroeconomic Policy", *Journal of Economic Literature*, 31, 105–141.

——— (1995): "Market Power in Macroeconomic Models: New Developments", *Annales d'Économie et de Statistique*, 37/38, 319–356.

TIROLE, J. (1989): *The Theory of Industrial Organization*. The MIT Press, Cambridge, MA.

TOBIN, J. (1993): "Price Flexibility and Output Stability: An Old Keynesian View", *Journal of Economic Perspectives*, 7, 45–65.

WEDDEPOHL, C. (1996): "Price Adjustment Models in a Production Economy", Report AE 11/96, University of Amsterdam.

WEIBULL, J. W. (1995): *Evolutionary Game Theory*. MIT Press, Cambridge, MA.

Lecture Notes in Economics and Mathematical Systems

For information about Vols. 1–290
please contact your bookseller or Springer-Verlag

Vol. 291: N. Takahashi, Design of Adaptive Organizations. VI, 140 pages. 1987.

Vol. 292: I. Tchijov, L. Tomaszewicz (Eds.), Input-Output Modeling. Proceedings, 1985. VI, 195 pages. 1987.

Vol. 293: D. Batten, J. Casti, B. Johansson (Eds.), Economic Evolution and Structural Adjustment. Proceedings, 1985. VI, 382 pages.

Vol. 294: J. Jahn, W. Knabs (Eds.), Recent Advances and Historical Development of Vector Optimization. VII, 405 pages. 1987.

Vol. 295. H. Meister, The Purification Problem for Constrained Games with Incomplete Information. X, 127 pages. 1987.

Vol. 296: A. Börsch-Supan, Econometric Analysis of Discrete Choice. VIII, 211 pages. 1987.

Vol. 297: V. Fedorov, H. Läuter (Eds.), Model-Oriented Data Analysis. Proceedings, 1987. VI, 239 pages. 1988.

Vol. 298: S.H. Chew, Q. Zheng, Integral Global Optimization. VII, 179 pages. 1988.

Vol. 299: K. Marti, Descent Directions and Efficient Solutions in Discretely Distributed Stochastic Programs. XIV, 178 pages. 1988.

Vol. 300. U. Derigs, Programming in Networks and Graphs. XI, 315 pages. 1988.

Vol. 301: J. Kacprzyk, M. Roubens (Eds.), Non-Conventional Preference Relations in Decision Making. VII, 155 pages. 1988.

Vol. 302: H.A. Eiselt, G. Pederzoli (Eds.), Advances in Optimization and Control. Proceedings, 1986. VIII, 372 pages. 1988.

Vol. 303: F.X. Diebold, Empirical Modeling of Exchange Rate Dynamics. VII, 143 pages. 1988.

Vol. 304: A. Kurzhanski, K. Neumann, D. Pallaschke (Eds.), Optimization, Parallel Processing and Applications. Proceedings, 1987. VI, 292 pages. 1988.

Vol. 305: G.-J.C.Th. van Schijndel, Dynamic Firm and Investor Behaviour under Progressive Personal Taxation. X, 215 pages. 1988.

Vol. 306: Ch. Klein, A Static Microeconomic Model of Pure Competition. VIII, 139 pages. 1988.

Vol. 307: T.K. Dijkstra (Ed.), On Model Uncertainty and its Statistical Implications. VII, 138 pages. 1988.

Vol. 308: J.R. Daduna, A. Wren (Eds.), Computer-Aided Transit Scheduling. VIII, 339 pages. 1988.

Vol. 309: G. Ricci, K. Velupillai (Eds.), Growth Cycles and Multisectoral Economics: The Goodwin Tradition. III, 126 pages. 1988.

Vol. 310: J. Kacprzyk, M. Fedrizzi (Eds.), Combining Fuzzy Imprecision with Probabilistic Uncertainty in Decision Making. IX, 399 pages. 1988.

Vol. 311: R. Färe, Fundamentals of Production Theory. IX, 163 pages. 1988.

Vol. 312: J. Krishnakumar, Estimation of Simultaneous Equation Models with Error Components Structure. X, 357 pages. 1988.

Vol. 313: W. Jammernegg, Sequential Binary Investment Decisions. VI, 156 pages. 1988.

Vol. 314: R. Tietz, W. Albers, R. Selten (Eds.), Bounded Rational Behavior in Experimental Games and Markets. VI, 368 pages. 1988.

Vol. 315: I. Orishimo, G.J.D. Hewings, P. Nijkamp (Eds), Information Technology: Social and Spatial Perspectives. Proceedings 1986. VI, 268 pages. 1988.

Vol. 316: R.L. Basmann, D.J. Slottje, K. Hayes, J.D. Johnson, D.J. Molina, The Generalized Fechner-Thurstone Direct Utility Function and Some of its Uses. VIII, 159 pages. 1988.

Vol. 317: L. Bianco, A. La Bella (Eds.), Freight Transport Planning and Logistics. Proceedings, 1987. X, 568 pages. 1988.

Vol. 318: T. Doup, Simplicial Algorithms on the Simplotope. VIII, 262 pages. 1988.

Vol. 319: D.T. Luc, Theory of Vector Optimization. VIII, 173 pages. 1989.

Vol. 320: D. van der Wijst, Financial Structure in Small Business. VII, 181 pages. 1989.

Vol. 321: M. Di Matteo, R.M. Goodwin, A. Vercelli (Eds.), Technological and Social Factors in Long Term Fluctuations. Proceedings. IX, 442 pages. 1989.

Vol. 322: T. Kollintzas (Ed.), The Rational Expectations Equilibrium Inventory Model. XI, 269 pages. 1989.

Vol. 323: M.B.M. de Koster, Capacity Oriented Analysis and Design of Production Systems. XII, 245 pages. 1989.

Vol. 324: I.M. Bomze, B.M. Pötscher, Game Theoretical Foundations of Evolutionary Stability. VI, 145 pages. 1989.

Vol. 325: P. Ferri, E. Greenberg, The Labor Market and Business Cycle Theories. X, 183 pages. 1989.

Vol. 326: Ch. Sauer, Alternative Theories of Output, Unemployment, and Inflation in Germany: 1960–1985. XIII, 206 pages. 1989.

Vol. 327: M. Tawada, Production Structure and International Trade. V, 132 pages. 1989.

Vol. 328: W. Güth, B. Kalkofen, Unique Solutions for Strategic Games. VII, 200 pages. 1989.

Vol. 329: G. Tillmann, Equity, Incentives, and Taxation. VI, 132 pages. 1989.

Vol. 330: P.M. Kort, Optimal Dynamic Investment Policies of a Value Maximizing Firm. VII, 185 pages. 1989.

Vol. 331: A. Lewandowski, A.P. Wierzbicki (Eds.), Aspiration Based Decision Support Systems. X, 400 pages. 1989.

Vol. 332: T.R. Gulledge, Jr., L.A. Litteral (Eds.), Cost Analysis Applications of Economics and Operations Research. Proceedings. VII, 422 pages. 1989.

Vol. 333: N. Dellaert, Production to Order. VII, 158 pages. 1989.

Vol. 334: H.-W. Lorenz, Nonlinear Dynamical Economics and Chaotic Motion. XI, 248 pages. 1989.

Vol. 335: A.G. Lockett, G. Islei (Eds.), Improving Decision Making in Organisations. Proceedings. IX, 606 pages. 1989.

Vol. 336: T. Puu, Nonlinear Economic Dynamics. VII, 119 pages. 1989.

Vol. 337: A. Lewandowski, I. Stanchev (Eds.), Methodology and Software for Interactive Decision Support. VIII, 309 pages. 1989.

Vol. 338: J.K. Ho, R.P. Sundarraj, DECOMP: An Implementation of Dantzig-Wolfe Decomposition for Linear Programming. VI, 206 pages.

Vol. 339: J. Terceiro Lomba, Estimation of Dynamic Econometric Models with Errors in Variables. VIII, 116 pages. 1990.

Vol. 340: T. Vasko, R. Ayres, L. Fontvieille (Eds.), Life Cycles and Long Waves. XIV, 293 pages. 1990.

Vol. 341: G.R. Uhlich, Descriptive Theories of Bargaining. IX, 165 pages. 1990.

Vol. 342: K. Okuguchi, F. Szidarovszky, The Theory of Oligopoly with Multi-Product Firms. V, 167 pages. 1990.

Vol. 343: C. Chiarella, The Elements of a Nonlinear Theory of Economic Dynamics. IX, 149 pages. 1990.

Vol. 344: K. Neumann, Stochastic Project Networks. XI, 237 pages. 1990.

Vol. 345: A. Cambini, E. Castagnoli, L. Martein, P Mazzoleni, S. Schaible (Eds.), Generalized Convexity and Fractional Programming with Economic Applications. Proceedings, 1988. VII, 361 pages. 1990.

Vol. 346: R. von Randow (Ed.), Integer Programming and Related Areas. A Classified Bibliography 1984–1987. XIII, 514 pages. 1990.

Vol. 347: D. Ríos Insua, Sensitivity Analysis in Multi-objective Decision Making. XI, 193 pages. 1990.

Vol. 348: H. Störmer, Binary Functions and their Applications. VIII, 151 pages. 1990.

Vol. 349: G.A. Pfann, Dynamic Modelling of Stochastic Demand for Manufacturing Employment. VI, 158 pages. 1990.

Vol. 350: W.-B. Zhang, Economic Dynamics. X, 232 pages. 1990.

Vol. 351: A. Lewandowski, V. Volkovich (Eds.), Multiobjective Problems of Mathematical Programming. Proceedings, 1988. VII, 315 pages. 1991.

Vol. 352: O. van Hilten, Optimal Firm Behaviour in the Context of Technological Progress and a Business Cycle. XII, 229 pages. 1991.

Vol. 353: G. Ricci (Ed.), Decision Processes in Economics. Proceedings, 1989. III, 209 pages 1991.

Vol. 354: M. Ivaldi, A Structural Analysis of Expectation Formation. XII, 230 pages. 1991.

Vol. 355: M. Salomon. Deterministic Lotsizing Models for Production Planning. VII, 158 pages. 1991.

Vol. 356: P. Korhonen, A. Lewandowski, J . Wallenius (Eds.), Multiple Criteria Decision Support. Proceedings, 1989. XII, 393 pages. 1991.

Vol. 357: P. Zörnig, Degeneracy Graphs and Simplex Cycling. XV, 194 pages. 1991.

Vol. 358: P. Knottnerus, Linear Models with Correlated Disturbances. VIII, 196 pages. 1991.

Vol. 359: E. de Jong, Exchange Rate Determination and Optimal Economic Policy Under Various Exchange Rate Regimes. VII, 270 pages. 1991.

Vol. 360: P. Stalder, Regime Translations, Spillovers and Buffer Stocks. VI, 193 pages . 1991.

Vol. 361: C. F. Daganzo, Logistics Systems Analysis. X, 321 pages. 1991.

Vol. 362: F. Gehrels, Essays in Macroeconomics of an Open Economy. VII, 183 pages. 1991.

Vol. 363: C. Puppe, Distorted Probabilities and Choice under Risk. VIII, 100 pages . 1991

Vol. 364: B. Horvath, Are Policy Variables Exogenous? XII, 162 pages. 1991.

Vol. 365: G. A. Heuer, U. Leopold-Wildburger. Balanced Silverman Games on General Discrete Sets. V, 140 pages. 1991.

Vol. 366: J. Gruber (Ed.), Econometric Decision Models. Proceedings, 1989. VIII, 636 pages. 1991.

Vol. 367: M. Grauer, D. B. Pressmar (Eds.), Parallel Computing and Mathematical Optimization. Proceedings. V, 208 pages. 1991.

Vol. 368: M. Fedrizzi, J. Kacprzyk, M. Roubens (Eds.), Interactive Fuzzy Optimization. VII, 216 pages. 1991.

Vol. 369: R. Koblo, The Visible Hand. VIII, 131 pages. 1991.

Vol. 370: M. J. Beckmann, M. N. Gopalan, R. Subramanian (Eds.), Stochastic Processes and their Applications. Proceedings, 1990. XLI, 292 pages. 1991.

Vol. 371: A. Schmutzler, Flexibility and Adjustment to Information in Sequential Decision Problems. VIII, 198 pages. 1991.

Vol. 372: J. Esteban, The Social Viability of Money. X, 202 pages. 1991.

Vol. 373: A. Billot, Economic Theory of Fuzzy Equilibria. XIII, 164 pages. 1992.

Vol. 374: G. Pflug, U. Dieter (Eds.), Simulation and Optimization. Proceedings, 1990. X, 162 pages. 1992.

Vol. 375: S.-J. Chen, Ch.-L. Hwang, Fuzzy Multiple Attribute Decision Making. XII, 536 pages. 1992.

Vol. 376: K.-H. Jöckel, G. Rothe, W. Sendler (Eds.), Bootstrapping and Related Techniques. Proceedings, 1990. VIII, 247 pages. 1992.

Vol. 377: A. Villar, Operator Theorems with Applications to Distributive Problems and Equilibrium Models. XVI, 160 pages. 1992.

Vol. 378: W. Krabs, J. Zowe (Eds.), Modern Methods of Optimization. Proceedings, 1990. VIII, 348 pages. 1992.

Vol. 379: K. Marti (Ed.), Stochastic Optimization. Proceedings, 1990. VII, 182 pages. 1992.

Vol. 380: J. Odelstad, Invariance and Structural Dependence. XII, 245 pages. 1992.

Vol. 381: C. Giannini, Topics in Structural VAR Econometrics. XI, 131 pages. 1992.

Vol. 382: W. Oettli, D. Pallaschke (Eds.), Advances in Optimization. Proceedings, 1991. X, 527 pages. 1992.

Vol. 383: J. Vartiainen, Capital Accumulation in a Corporatist Economy. VII, 177 pages. 1992.

Vol. 384: A. Martina, Lectures on the Economic Theory of Taxation. XII, 313 pages. 1992.

Vol. 385: J. Gardeazabal, M. Regúlez, The Monetary Model of Exchange Rates and Cointegration. X, 194 pages. 1992.

Vol. 386: M. Desrochers, J.-M. Rousseau (Eds.), Computer-Aided Transit Scheduling. Proceedings, 1990. XIII, 432 pages. 1992.

Vol. 387: W. Gaertner, M. Klemisch-Ahlert, Social Choice and Bargaining Perspectives on Distributive Justice. VIII, 131 pages. 1992.

Vol. 388: D. Bartmann, M. J. Beckmann, Inventory Control. XV, 252 pages. 1992.

Vol. 389: B. Dutta, D. Mookherjee, T. Parthasarathy, T. Raghavan, D. Ray, S. Tijs (Eds.), Game Theory and Economic Applications. Proceedings, 1990. IX, 454 pages. 1992.

Vol. 390: G. Sorger, Minimum Impatience Theorem for Recursive Economic Models. X, 162 pages. 1992.

Vol. 391: C. Keser, Experimental Duopoly Markets with Demand Inertia. X, 150 pages. 1992.

Vol. 392: K. Frauendorfer, Stochastic Two-Stage Programming. VIII, 228 pages. 1992.

Vol. 393: B. Lucke, Price Stabilization on World Agricultural Markets. XI, 274 pages. 1992.

Vol. 394: Y.-J. Lai, C.-L. Hwang, Fuzzy Mathematical Programming. XIII, 301 pages. 1992.

Vol. 395: G. Haag, U. Mueller, K. G. Troitzsch (Eds.), Economic Evolution and Demographic Change. XVI, 409 pages. 1992.

Vol. 396: R. V. V. Vidal (Ed.), Applied Simulated Annealing. VIII, 358 pages. 1992.

Vol. 397: J. Wessels, A. P. Wierzbicki (Eds.), User-Oriented Methodology and Techniques of Decision Analysis and Support. Proceedings, 1991. XII, 295 pages. 1993.

Vol. 398: J.-P. Urbain, Exogeneity in Error Correction Models. XI, 189 pages. 1993.

Vol. 399: F. Gori, L. Geronazzo, M. Galeotti (Eds.), Nonlinear Dynamics in Economics and Social Sciences. Proceedings, 1991. VIII, 367 pages. 1993.

Vol. 400: H. Tanizaki, Nonlinear Filters. XII, 203 pages. 1993.

Vol. 401: K. Mosler, M. Scarsini, Stochastic Orders and Applications. V, 379 pages. 1993.

Vol. 402: A. van den Elzen, Adjustment Processes for Exchange Economies and Noncooperative Games. VII, 146 pages. 1993.

Vol. 403: G. Brennscheidt, Predictive Behavior. VI, 227 pages. 1993.

Vol. 404: Y.-J. Lai, Ch.-L. Hwang, Fuzzy Multiple Objective Decision Making. XIV, 475 pages. 1994.

Vol. 405: S. Komlósi, T. Rapcsák, S. Schaible (Eds.), Generalized Convexity. Proceedings, 1992. VIII, 404 pages. 1994.

Vol. 406: N. M. Hung, N. V. Quyen, Dynamic Timing Decisions Under Uncertainty. X, 194 pages. 1994.

Vol. 407: M. Ooms, Empirical Vector Autoregressive Modeling. XIII, 380 pages. 1994.

Vol. 408: K. Haase, Lotsizing and Scheduling for Production Planning. VIII, 118 pages. 1994.

Vol. 409: A. Sprecher, Resource-Constrained Project Scheduling. XII, 142 pages. 1994.

Vol. 410: R. Winkelmann, Count Data Models. XI, 213 pages. 1994.

Vol. 411: S. Dauzère-Péres, J.-B. Lasserre, An Integrated Approach in Production Planning and Scheduling. XVI, 137 pages. 1994.

Vol. 412: B. Kuon, Two-Person Bargaining Experiments with Incomplete Information. IX, 293 pages. 1994.

Vol. 413: R. Fiorito (Ed.), Inventory, Business Cycles and Monetary Transmission. VI, 287 pages. 1994.

Vol. 414: Y. Crama, A. Oerlemans, F. Spieksma, Production Planning in Automated Manufacturing. X, 210 pages. 1994.

Vol. 415: P. C. Nicola, Imperfect General Equilibrium. XI, 167 pages. 1994.

Vol. 416: H. S. J. Cesar, Control and Game Models of the Greenhouse Effect. XI, 225 pages. 1994.

Vol. 417: B. Ran, D. E. Boyce, Dynamic Urban Transportation Network Models. XV, 391 pages. 1994.

Vol. 418: P. Bogetoft, Non-Cooperative Planning Theory. XI, 309 pages. 1994.

Vol. 419: T. Maruyama, W. Takahashi (Eds.), Nonlinear and Convex Analysis in Economic Theory. VIII, 306 pages. 1995.

Vol. 420: M. Peeters, Time-To-Build. Interrelated Investment and Labour Demand Modelling. With Applications to Six OECD Countries. IX, 204 pages. 1995.

Vol. 421: C. Dang, Triangulations and Simplicial Methods. IX, 196 pages. 1995.

Vol. 422: D. S. Bridges, G. B. Mehta, Representations of Preference Orderings. X, 165 pages. 1995.

Vol. 423: K. Marti, P. Kall (Eds.), Stochastic Programming. Numerical Techniques and Engineering Applications. VIII, 351 pages. 1995.

Vol. 424: G. A. Heuer, U. Leopold-Wildburger, Silverman's Game. X, 283 pages. 1995.

Vol. 425: J. Kohlas, P.-A. Monney, A Mathematical Theory of Hints. XIII, 419 pages, 1995.

Vol. 426: B. Finkenstädt, Nonlinear Dynamics in Economics. IX, 156 pages. 1995.

Vol. 427: F. W. van Tongeren, Microsimulation Modelling of the Corporate Firm. XVII, 275 pages. 1995.

Vol. 428: A. A. Powell, Ch. W. Murphy, Inside a Modern Macroeconometric Model. XVIII, 424 pages. 1995.

Vol. 429: R. Durier, C. Michelot, Recent Developments in Optimization. VIII, 356 pages. 1995.

Vol. 430: J. R. Daduna, I. Branco, J. M. Pinto Paixão (Eds.), Computer-Aided Transit Scheduling. XIV, 374 pages. 1995.

Vol. 431: A. Aulin, Causal and Stochastic Elements in Business Cycles. XI, 116 pages. 1996.

Vol. 432: M. Tamiz (Ed.), Multi-Objective Programming and Goal Programming. VI, 359 pages. 1996.

Vol. 433: J. Menon, Exchange Rates and Prices. XIV, 313 pages. 1996.

Vol. 434: M. W. J. Blok, Dynamic Models of the Firm. VII, 193 pages. 1996.

Vol. 435: L. Chen, Interest Rate Dynamics, Derivatives Pricing, and Risk Management. XII, 149 pages. 1996.

Vol. 436: M. Klemisch-Ahlert, Bargaining in Economic and Ethical Environments. IX, 155 pages. 1996.

Vol. 437: C. Jordan, Batching and Scheduling. IX, 178 pages. 1996.

Vol. 438: A. Villar, General Equilibrium with Increasing Returns. XIII, 164 pages. 1996.

Vol. 439: M. Zenner, Learning to Become Rational. VII, 201 pages. 1996.

Vol. 440: W. Ryll, Litigation and Settlement in a Game with Incomplete Information. VIII, 174 pages. 1996.

Vol. 441: H. Dawid, Adaptive Learning by Genetic Algorithms. IX, 166 pages.1996.

Vol. 442: L. Corchón, Theories of Imperfectly Competitive Markets. XIII, 163 pages. 1996.

Vol. 443: G. Lang, On Overlapping Generations Models with Productive Capital. X, 98 pages. 1996.

Vol. 444: S. Jørgensen, G. Zaccour (Eds.), Dynamic Competitive Analysis in Marketing. X, 285 pages. 1996.

Vol. 445: A. H. Christer, S. Osaki, L. C. Thomas (Eds.), Stochastic Modelling in Innovative Manufactoring. X, 361 pages. 1997.

Vol. 446: G. Dhaene, Encompassing. X, 160 pages. 1997.

Vol. 447: A. Artale, Rings in Auctions. X, 172 pages. 1997.

Vol. 448: G. Fandel, T. Gal (Eds.), Multiple Criteria Decision Making. XII, 678 pages. 1997.

Vol. 449: F. Fang, M. Sanglier (Eds.), Complexity and Self-Organization in Social and Economic Systems. IX, 317 pages, 1997.

Vol. 450: P. M. Pardalos, D. W. Hearn, W. W. Hager, (Eds.), Network Optimization. VIII, 485 pages, 1997.

Vol. 451: M. Salge, Rational Bubbles. Theoretical Basis, Economic Relevance, and Empirical Evidence with a Special Emphasis on the German Stock Market.IX, 265 pages. 1997.

Vol. 452: P. Gritzmann, R. Horst, E. Sachs, R. Tichatschke (Eds.), Recent Advances in Optimization. VIII, 379 pages. 1997.

Vol. 453: A. S. Tangian, J. Gruber (Eds.), Constructing Scalar-Valued Objective Functions. VIII, 298 pages. 1997.

Vol. 454: H.-M. Krolzig, Markov-Switching Vector Autoregressions. XIV, 358 pages. 1997.

Vol. 455: R. Caballero, F. Ruiz, R. E. Steuer (Eds.), Advances in Multiple Objective and Goal Programming. VIII, 391 pages. 1997.

Vol. 456: R. Conte, R. Hegselmann, P. Terna (Eds.), Simulating Social Phenomena. VIII, 536 pages. 1997.

Vol. 457: C. Hsu, Volume and the Nonlinear Dynamics of Stock Returns. VIII, 133 pages. 1998.

Vol. 458: K. Marti, P. Kall (Eds.), Stochastic Programming Methods and Technical Applications. X, 437 pages. 1998.

Vol. 459: H. K. Ryu, D. J. Slottje, Measuring Trends in U.S. Income Inequality. XI, 195 pages. 1998.

Vol. 460: B. Fleischmann, J. A. E. E. van Nunen, M. G. Speranza, P. Stähly, Advances in Distribution Logistic. XI, 535 pages. 1998.

Vol. 461: U. Schmidt, Axiomatic Utility Theory under Risk. XV, 201 pages. 1998.

Vol. 462: L. von Auer, Dynamic Preferences, Choice Mechanisms, and Welfare. XII, 226 pages. 1998.

Vol. 463: G. Abraham-Frois (Ed.), Non-Linear Dynamics and Endogenous Cycles. VI, 204 pages. 1998.

Vol. 464: A. Aulin, The Impact of Science on Economic Growth and its Cycles. IX, 204 pages. 1998.

Vol. 465: T. J. Stewart, R. C. van den Honert (Eds.), Trends in Multicriteria Decision Making. X, 448 pages. 1998.

Vol. 466: A. Sadrieh, The Alternating Double Auction Market. VII, 350 pages. 1998.

Vol. 467: H. Hennig-Schmidt, Bargaining in a Video Experiment. Determinants of Boundedly Rational Behavior. XII, 221 pages. 1999.

Vol. 468: A. Ziegler, A Game Theory Analysis of Options. XIV, 145 pages. 1999.

Vol. 469: M. P. Vogel, Environmental Kuznets Curves. XIII, 197 pages. 1999.

Vol. 470: M. Ammann, Pricing Derivative Credit Risk. XII, 228 pages. 1999.

Vol. 471: N. H. M. Wilson (Ed.), Computer-Aided Transit Scheduling. XI, 444 pages. 1999.

Vol. 472: J.-R. Tyran, Money Illusion and Strategic Complementarity as Causes of Monetary Non-Neutrality. X, 228 pages. 1999.

Vol. 473: S. Helber, Performance Analysis of Flow Lines with Non-Linear Flow of Material. IX, 280 pages. 1999.

Vol. 474: U. Schwalbe, The Core of Economies with Asymmetric Information. IX, 141 pages. 1999.

Vol. 475: L. Kaas, Dynamic Macroelectronics with Imperfect Competition. XI, 155 pages. 1999.

GPSR Compliance

The European Union's (EU) General Product Safety Regulation (GPSR) is a set of rules that requires consumer products to be safe and our obligations to ensure this.

If you have any concerns about our products, you can contact us on ProductSafety@springernature.com

In case Publisher is established outside the EU, the EU authorized representative is:

Springer Nature Customer Service Center GmbH
Europaplatz 3
69115 Heidelberg, Germany

Batch number: 10844675

Printed by Printforce, the Netherlands